Preface

SPSS, standing for *Statistical Package for the Social Sciences*, is a powerful, user-friendly software package for the manipulation and statistical analysis of data. The package is particularly useful for students and researchers in psychology, sociology, psychiatry, and other behavioral sciences, containing as it does an extensive range of both univariate and multivariate procedures much used in these disciplines. Our aim in this handbook is to give brief and straightforward descriptions of how to conduct a range of statistical analyses using the latest version of SPSS, SPSS 11. Each chapter deals with a different type of analytical procedure applied to one or more data sets primarily (although not exclusively) from the social and behavioral areas. Although we concentrate largely on how to use SPSS to get results and on how to correctly interpret these results, the basic theoretical background of many of the techniques used is also described in separate boxes. When more advanced procedures are used, readers are referred to other sources for details. Many of the boxes contain a few mathematical formulae, but by separating this material from the body of the text, we hope that even readers who have limited mathematical background will still be able to undertake appropriate analyses of their data.

The text is not intended in any way to be an introduction to statistics and, indeed, we assume that most readers will have attended at least one statistics course and will be relatively familiar with concepts such as *linear regression, correlation, significance tests,* and simple *analysis of variance.* Our hope is that researchers and students with such a background will find this book a relatively self-contained means of using SPSS to analyze their data correctly.

Each chapter ends with a number of exercises, some relating to the data sets introduced in the chapter and others introducing further data sets. Working through these exercises will develop both SPSS and statistical skills. Answers to most of the exercises in the text are provided at

v

http://www.iop.kcl.ac.uk/iop/departments/BioComp/SPSSBook.shtml. The majority of data sets used in the book can be found at the same site.

We are grateful to Ms. Harriet Meteyard for her usual excellent word processing and overall support during the writing of this book.

Sabine Landau and Brian Everitt
London, July 2003

A Handbook of
Statistical Analyses
using SPSS

A Handbook of Statistical Analyses using SPSS

Sabine Landau

and

Brian S. Everitt

CHAPMAN & HALL/CRC

A CRC Press Company

Boca Raton   London   New York   Washington, D.C.

Library of Congress Cataloging-in-Publication Data

Landau, Sabine.
A handbook of statistical analyses using SPSS / Sabine, Landau, Brian S. Everitt.
p. cm.
Includes bibliographical references and index.
ISBN 1-58488-369-3 (alk. paper)
1. SPSS (Computer file). 2. Social sciences—Statistical methods—Computer programs. 3. Social sciences—Statistical methods—Data processing. I. Everitt, Brian S. II. Title.

HA32.E93 2003
519.5′0285—dc22   2003058474

Visit the CRC Press Web site at www.crcpress.com

© 2004 by Chapman & Hall/CRC CRC Press LLC

No claim to original U.S. Government works
International Standard Book Number 1-58488-369-3
Library of Congress Card Number 2003058474
Printed in the United States of America 1 2 3 4 5 6 7 8 9 0
Printed on acid-free paper

Distributors

The distributor for SPSS in the United Kingdom is

SPSS U.K. Ltd.
1st Floor St. Andrew's House, West Street
Woking
Surrey, United Kingdom GU21 6EB
Tel.   0845 3450935
FAX   01483 719290
Email   sales@spss.co.uk

In the United States, the distributor is

SPSS Inc.
233 S. Wacker Drive, 11th floor
Chicago, IL 60606-6307
Tel.   1(800) 543-2185
FAX   1(800) 841-0064
Email   sales@spss.com

Contents

Chapter 1

A Brief Introduction to SPSS

1.1 Introduction

The "Statistical Package for the Social Sciences" (SPSS) is a package of programs for manipulating, analyzing, and presenting data; the package is widely used in the social and behavioral sciences. There are several forms of SPSS. The core program is called *SPSS Base* and there are a number of add-on modules that extend the range of data entry, statistical, or reporting capabilities. In our experience, the most important of these for statistical analysis are the *SPSS Advanced Models* and *SPSS Regression Models* add-on modules. SPSS Inc. also distributes stand-alone programs that work with SPSS.

There are versions of SPSS for Windows (98, 2000, ME, NT, XP), major UNIX platforms (Solaris, Linux, AIX), and Macintosh. In this book, we describe the most popular, SPSS for Windows, although most features are shared by the other versions. The analyses reported in this book are based on SPSS version 11.0.1 running under Windows 2000. By the time this book is published, there will almost certainly be later versions of SPSS available, but we are confident that the SPSS instructions given in each of the chapters will remain appropriate for the analyses described.

While writing this book we have used the *SPSS Base, Advanced Models, Regression Models,* and the *SPSS Exact Tests* add-on modules. Other available add-on modules (*SPSS Tables, SPSS Categories, SPSS Trends, SPSS Missing Value Analysis*) were not used.

1. *SPSS Base* (Manual: *SPSS Base 11.0 for Windows User's Guide):* This provides methods for data description, simple inference for continuous and categorical data and linear regression and is, therefore, sufficient to carry out the analyses in Chapters 2, 3, and 4. It also provides techniques for the analysis of multivariate data, specifically for factor analysis, cluster analysis, and discriminant analysis (see Chapters 11 and 12).

2. *Advanced Models module* (Manual: *SPSS 11.0 Advanced Models*): This includes methods for fitting general linear models and linear mixed models and for assessing survival data, and is needed to carry out the analyses in Chapters 5 through 8 and in Chapter 10.

3. *Regression Models module* (Manual: *SPSS 11.0 Regression Models*): This is applicable when fitting nonlinear regression models. We have used it to carry out a logistic regression analysis (see Chapter 9).

(*The Exact Tests* module has also been employed on occasion, specifically in the Exercises for Chapters 2 and 3, to generate exact p-values.)

The *SPSS 11.0 Syntax Reference Guide* (SPSS, Inc., 2001c) is a reference for the command syntax for the SPSS Base system and the Regression Models and Advanced Models options.

The SPSS Web site (http://www.spss.com/) provides information on add-on modules and stand-alone packages working with SPSS, events and SPSS user groups. It also supplies technical reports and maintains a frequently asked questions (FAQs) list.

SPSS for Windows offers a spreadsheet facility for entering and browsing the working data file — the **Data Editor**. Output from statistical procedures is displayed in a separate window — the **Output Viewer**. It takes the form of tables and graphics that can be manipulated interactively and can be copied directly into other applications.

It is its *graphical user interface* (GUI) that makes SPSS so easy by simply selecting procedures from the many menus available. It is the GUI that is used in this book to carry out all the statistical analysis presented. We also show how to produce command syntax for record keeping.

We assume that the reader is already familiar with the Windows GUI and we do not spend much time discussing the data manipulation and result presentation facilities of SPSS for Windows. These features are described in detail in the *Base User's Guide* (SPSS, Inc., 2001d). Rather we focus on the statistical features of SPSS — showing how it can be used to carry out statistical analyses of a variety of data sets and on how to interpret the resulting output. To aid in reading this text, we have adopted the **Helvetica Narrow** font to indicate spreadsheet column names, menu commands, and text in dialogue boxes as seen on the SPSS GUI.

1.2 Getting Help

Online help is provided from the **Help** menu or via context menus or **Help** buttons on dialogue boxes. We will mention the latter features when discussing the dialogue boxes and output tables. Here, we concentrate on the general help facility. The required menu is available from any window and provides three major help facilities:

Help — Statistics Coach helps users unfamiliar with SPSS or the statistical procedures available in SPSS to get started. This facility prompts the user with simple questions in nontechnical language about the purpose of the statistical analysis and provides visual examples of basic statistical and charting features in SPSS. The facility covers only a selected subset of procedures.

Help — Tutorial provides access to an introductory SPSS tutorial, including a comprehensive overview of SPSS basics. It is designed to provide a step-by-step guide for carrying out a statistical analysis in SPSS. All files shown in the examples are installed with the tutorial so the user can repeat the analysis steps.

Help — Topics opens the **Help Topics: SPSS for Windows** box, which provides access to **Contents, Index,** and **Find** tabs. Under the **Contents** tab, double-clicking items with a book symbol expands or collapses their contents (the **Open** and **Close** buttons do the same). The **Index** tab provides an alphabetical list of topics. Once a topic is selected (by double-clicking), or the first few letters of the word are typed in, the **Display** button provides a description. The **Find** tab allows for searching the help files for specific words and phrases.

1.3 Data Entry

When SPSS 11.0 for Windows is first opened, a default dialogue box appears that gives the user a number of options. The **Tutorial** can be accessed at this stage. Most likely users will want to enter data or open an existing data file; we demonstrate the former (Display 1.1). Further options will be discussed later in this chapter. This dialogue box can be prevented from opening in the future by checking this option at the bottom of the box.

When **Type in data** is selected, the SPSS **Data Editor** appears as an empty spreadsheet. At the top of the screen is a menu bar and at the bottom a status bar. The status bar informs the user about facilities currently active; at the beginning of a session it simply reads, "SPSS Processor is ready."

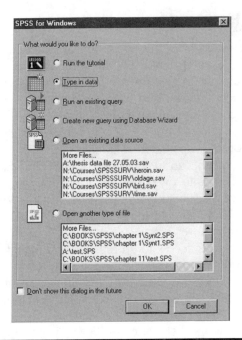

Display 1.1 Initial SPSS for Windows **dialogue box.**

The facilities provided by the menus will be explained later in this chapter. SPSS also provides a toolbar for quick and easy access to common tasks. A brief description of each tool can be obtained by placing the cursor over the tool symbol and the display of the toolbar can be controlled using the command **Toolbars...** from the **View** drop-down menu (for more details, see the *Base User's Guide*, SPSS Inc., 2001d).

1.3.1 The Data View Spreadsheet

The **Data Editor** consists of two windows. By default the **Data View**, which allows the data to be entered and viewed, is shown (Display 1.2). The other window is the **Variable View**, which allows the types of variables to be specified and viewed. The user can toggle between the windows by clicking on the appropriate tabs on the bottom left of the screen.

Data values can be entered in the **Data View** spreadsheet. For most analysis SPSS assumes that rows represent cases and columns variables. For example, in Display 1.2 some of five available variable values have been entered for twenty subjects. By default SPSS aligns numerical data entries to the right-hand side of the cells and text (string) entries to the left-hand side. Here variables **sex**, **age**, **extrover**, and **car** take numerical

Display 1.2   Data View **window of the** Data Editor.

values while the variable **make** takes string values. By default SPSS uses
a period/full stop to indicate missing numerical values. String variable
cells are simply left empty. Here, for example, the data for variables
extrover, car, and **make** have not yet been typed in for the 20 subjects so
the respective values appear as missing.

The appearance of the **Data View** spreadsheet is controlled by the **View**
drop-down menu. This can be used to change the font in the cells, remove
lines, and make value labels visible. When labels have been assigned to
the category codes of a categorical variable, these can be displayed by
checking **Value Labels** (or by selecting ![icon] on the toolbar). Once the category
labels are visible, highlighting a cell produces a button with a downward
arrow on the right-hand side of the cell. Clicking on this arrow produces
a drop-down list with all the available category labels for the variable.
Clicking on any of these labels results in the respective category and label
being inserted in the cell. This feature is useful for editing the data.

1.3.2 The Variable View Spreadsheet

The **Variable View** spreadsheet serves to define the variables (Display 1.3).
Each variable definition occupies a row of this spreadsheet. As soon as
data is entered under a column in the **Data View**, the default name of the
column occupies a row in the **Variable View**.

Display 1.3 Variable View **window of the** Data Editor.

There are 10 characteristics to be specified under the columns of the Variable View (Display 1.3):

1. **Name** — the chosen variable name. This can be up to eight alphanumeric characters but must begin with a letter. While the underscore (_) is allowed, hyphens (-), ampersands (&), and spaces cannot be used. Variable names are not case sensitive.

2. **Type** — the type of data. SPSS provides a default variable type once variable values have been entered in a column of the Data View. The type can be changed by highlighting the respective entry in the second column of the Variable View and clicking the three-periods symbol (…) appearing on the right-hand side of the cell. This results in the Variable Type box being opened, which offers a number of types of data including various formats for numerical data, dates, or currencies. (Note that a common mistake made by first-time users is to enter categorical variables as type "string" by typing text into the Data View. To enable later analyses, categories should be given artificial number codes and defined to be of type "numeric.")

3. **Width** — the width of the actual data entries. The default width of numerical variable entries is eight. The width can be increased or decreased by highlighting the respective cell in the third column and employing the upward or downward arrows appearing on the

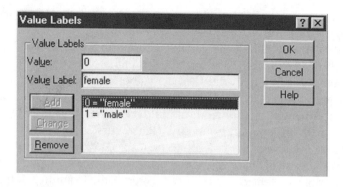

Display 1.4   Declaring category code labels.

right-hand side of the cell or by simply typing a new number in
the cell.

4. **Decimals** — the number of digits to the right of the decimal place
   to be displayed for data entries. This is not relevant for string data
   and for such variables the entry under the fourth column is given
   as a greyed-out zero. The value can be altered in the same way
   as the value of **Width**.

5. **Label** — a label attached to the variable name. In contrast to the
   variable name, this is not confined to eight characters and spaces
   can be used. It is generally a good idea to assign variable labels.
   They are helpful for reminding users of the meaning of variables
   (placing the cursor over the variable name in the **Data View** will
   make the variable label appear) and can be displayed in the output
   from statistical analyses.

6. **Values** — labels attached to category codes. For categorical variables,
   an integer code should be assigned to each category and the
   variable defined to be of type "numeric." When this has been done,
   clicking on the respective cell under the sixth column of the **Variable
   View** makes the three-periods symbol appear, and clicking this
   opens the **Value Labels** dialogue box, which in turn allows assignment of labels to category codes. For example, our data set included
   a categorical variable **sex** indicating the gender of the subject.
   Clicking the three-periods symbol opens the dialogue box shown
   in Display 1.4 where numerical code "0" was declared to represent
   females and code "1" males.

7. **Missing** — missing value codes. SPSS recognizes the period symbol
   as indicating a missing value. If other codes have been used (e.g.,
   99, 999) these have to be declared to represent missing values by
   highlighting the respective cell in the seventh column, clicking the

three-periods symbol and filling in the resulting Missing Values dialogue box accordingly.

8. Columns — width of the variable column in the Data View. The default
   cell width for numerical variables is eight. Note that when the Width
   value is larger than the Columns value, only part of the data entry
   might be seen in the Data View. The cell width can be changed in
   the same way as the width of the data entries or simply by dragging
   the relevant column boundary. (Place cursor on right-hand boundary of the title of the column to be resized. When the cursor changes
   into a vertical line with a right and left arrow, drag the cursor to
   the right or left to increase or decrease the column width.)

9. Align — alignment of variable entries. The SPSS default is to align
   numerical variables to the right-hand side of a cell and string
   variables to the left. It is generally helpful to adhere to this default;
   but if necessary, alignment can be changed by highlighting the
   relevant cell in the ninth column and choosing an option from the
   drop-down list.

10. Measure — measurement scale of the variable. The default chosen
    by SPSS depends on the data type. For example, for variables of
    type "numeric," the default measurement scale is a continuous or
    interval scale (referred to by SPSS as "scale"). For variables of type
    "string," the default is a nominal scale. The third option, "ordinal,"
    is for categorical variables with ordered categories but is not used
    by default. It is good practice to assign each variable the highest
    appropriate measurement scale ("scale" > "ordinal" > "nominal")
    since this has implications for the statistical methods that are
    applicable. The default setting can be changed by highlighting the
    respective cell in the tenth column and choosing an appropriate
    option from the drop-down list.

A summary of variable characteristics can be obtained from the Utilities
drop-down menu. The Variables... command opens a dialogue box where
information can be requested for a selected variable, while choosing File
Info from the drop-down menu generates this information for every variable
in the Data View.

1.4 Storing and Retrieving Data Files

Storing and retrieving data files are carried out via the drop-down menu
available after selecting File on the menu bar (Display 1.5).

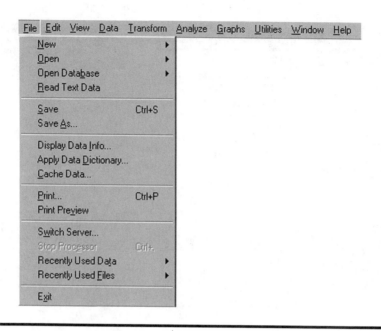

Display 1.5 File **drop-down menu.**

A data file shown in the **Data Editor** can be saved by using the commands **Save** or **Save As**.... In the usual Windows fashion **Save** (or ■ from the toolbar) will save the data file under its current name, overwriting an existing file or prompting for a name otherwise. By contrast, **Save As** always opens the **Save Data As** dialogue where the directory, file name, and type have to be specified. SPSS supports a number of data formats. SPSS data files are given the extension *.sav. Other formats, such as ASCII text (*.dat), Excel (*.xls), or dBASE (*.dbf), are also available.

To open existing SPSS data files we use the commands **File – Open – Data...** from the menu bar (or ☞ from the toolbar). This opens the **Open File** dialogue box from which the appropriate file can be chosen in the usual way (Display 1.6). Recently used files are also accessible by placing the cursor over **Recently Used Data** on the **File** drop-down menu and double-clicking on the required file. In addition, files can be opened when first starting SPSS by checking **Open an existing data source** on the initial dialogue box (see Display 1.1).

SPSS can import data files in other than SPSS format. A list of data formats is provided by selecting the down arrow next to the **Files of type** field (Display 1.6). There are a number of formats including spreadsheet (e.g., Excel, *.xls), database (e.g., dBase, *.dbf), and ACSII text (e.g., *.txt,

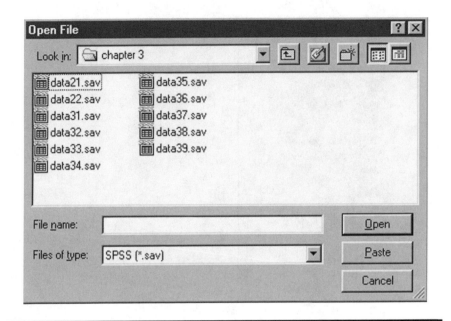

Display 1.6   Opening an existing SPSS data file.

*.dat). Selecting a particular file extension will cause a dialogue box to appear that asks for information relevant to the format. Here we briefly discuss importing Excel files and ASCII text files.

Selecting to import an Excel spreadsheet in the **Open File** box will bring up the **Opening File Options** box. If the spreadsheet contains a row with variable names, **Read Variable Names** has to be checked in this box in order that the first row of the spreadsheet is read into variable names. In addition, if there are initial empty rows or columns in the spreadsheet, SPSS needs to be informed about it by defining the cells to be read in the **Range** field of the **Opening File Options** box (using the standard spreadsheet format, e.g., B4:H10 for the cells in the rectangle with corners B4 and H10 inclusive).

Selecting to open an ASCII text file in the **Open File** dialogue box (or selecting **Read Text Data** from the **File** drop-down directly, see Display 1.5) causes the **Text Import Wizard** to start. The initial dialogue box is shown in Display 1.7. The **Wizard** proceeds in six steps asking questions related to the import format (e.g., how the variables are arranged, whether variable names are included in the text file), while at the same time making suggestions and displaying how the text file would be transformed into an SPSS spreadsheet. The **Text Import Wizard** is a convenient and self-explanatory ASCII text import tool.

(Choosing **New** from the **File** drop-down menu will clear the **Data Editor** spreadsheet for entry of new data.)

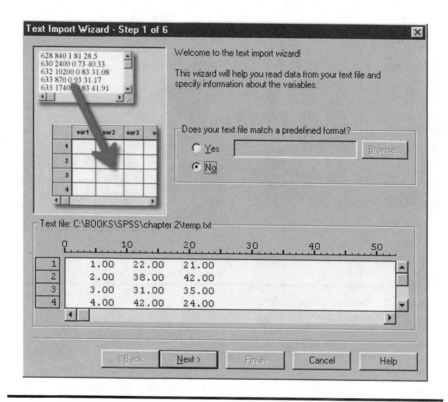

Display 1.7 **Text Import Wizard dialogue box.**

1.5 The Statistics Menus

The drop-down menus available after selecting Data, Transform, Analyze, or Graphs from the menu bar provide procedures concerned with different aspects of a statistical analysis. They allow manipulation of the format of the data spreadsheet to be used for analysis (Data), generation of new variables (Transform), running of statistical procedures (Analyze), and construction of graphical displays (Graphs).

Most statistics menu selections open dialogue boxes; a typical example is shown in Display 1.8. The dialogue boxes are used to select variables and options for analysis. A main dialogue for a statistical procedure has several components:

- ■ **A source variables list** is a list of variables from the Data View spreadsheet that can be used in the requested analysis. Only variable types that are allowed by the procedure are displayed in the source list. Variables of type "string" are often not allowed. A

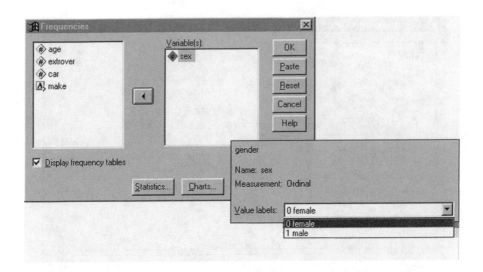

Display 1.8   Typical dialogue box.

sign icon next to the variable name indicates the variable type. A
hash sign (#) is used for numeric variables and "A" indicates that
the variable is a string variable.

■ **Target variable(s) lists** are lists indicating the variables to be
included in the analysis (e.g., lists of dependent and independent
variables).

■ **Command buttons** are buttons that can be pressed to instruct
the program to perform an action. For example, run the procedure
(click **OK**), reset all specifications to the default setting (click **Reset**),
display context sensitive help (click **Help**), or open a **sub-dialogue
box** for specifying additional procedure options.

Information about variables shown in a dialogue box can be obtained
by simply highlighting the variable by left-clicking and then right-clicking
and choosing **Variable Information** in the pop-up context menu. This results
in a display of the variable label, name, measurement scale, and value
labels if applicable (Display 1.8).

It is also possible to right-click on any of the controls or variables in
a dialogue box to obtain a short description. For controls, a description
is provided automatically after right-clicking. For variables, **What's this?** must
be chosen from the pop-up context menu.

SPSS provides a choice between displaying variable names or variable
labels in the dialogue boxes. While variable labels can provide more
accurate descriptions of the variables, they are often not fully displayed

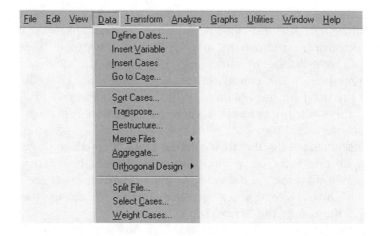

Display 1.9 Data **drop-down menu.**

in a box due to their length (positioning the cursor over the variable label will show the whole text). We, therefore, prefer to display variable names and have adhered to this setting in all the dialogue boxes shown later in this book. Displays are controlled via the **Options** dialogue box opened by using the commands, **Edit – Option...** from the menu bar. To display variable names, check **Display names** under **Variable Lists** on the **General** tab.

1.5.1 *Data File Handling*

The data file as displayed in the **Data View** spreadsheet is not always organized in the appropriate format for a particular use. The **Data** drop-down menu provides procedures for reorganizing the structure of a data file (Display 1.9).

The first four command options from the **Data** drop-down menu are concerned with editing or moving within the **Data View** spreadsheet. Date formats can be defined or variables or cases inserted.

The following set of procedures allows the format of a data file to be changed:

■ **Sort Cases...** opens a dialogue box that allows sorting of cases (rows) in the spreadsheet according to the values of one or more variables. Cases can be arranged in ascending or descending order. When several sorting variables are employed, cases will be sorted by each variable within categories of the prior variable on the **Sort by** list. Sorting can be useful for generating graphics (see an example of this in Chapter 10).

■ **Transpose...** opens a dialogue for swapping the rows and columns of the data spreadsheet. The **Variable(s)** list contains the columns to be transposed into rows and an ID variable can be supplied as the **Name Variable** to name the columns of the new transposed spreadsheet. The command can be useful when procedures normally used on the columns of the spreadsheet are to be applied to the rows, for example, to generate summary measures of case-wise repeated measures.

■ **Restructure...** calls the **Restructure Data Wizard**, a series of dialogue boxes for converting data spreadsheets between what are known as "long" and "wide" formats. These formats are relevant in the analysis of repeated measures and we will discuss the formats and the use of the **Wizard** in Chapter 8.

■ **Merge files** allows either **Add Cases...** or **Add Variables...** to an existing data file. A dialogue box appears that allows opening a second data file. This file can either contain the same variables but different cases (to add cases) or different variables but the same cases (to add variables). The specific requirements of these procedures are described in detail in the *Base User's Guide* (SPSS Inc., 2001d). The commands are useful at the database construction stage of a project and offer wide-ranging options for combining data files.

■ **Aggregate...** combines groups of rows (cases) into single summary rows and creates a new aggregated data file. The grouping variables are supplied under the **Break Variable(s)** list of the **Aggregate Data** dialogue box and the variables to be aggregated under the **Aggregate Variable(s)** list. The **Function...** sub-dialogue box allows for the aggregation function of each variable to be chosen from a number of possibilities (mean, median, value of first case, number of cases, etc.). This command is useful when the data are of a hierarchical structure, for example, patients within wards within hospitals. The data file might be aggregated when the analysis of characteristics of higher level units (e.g., wards, hospitals) is of interest.

Finally, the **Split File...**, **Select Cases...**, and **Weight Cases...** procedures allow using the data file in a particular format without changing its appearance in the **Data View** spreadsheet. These commands are frequently used in practical data analysis and we provide several examples in later chapters. Here we will describe the **Select Cases...** and **Split File...** commands. The **Weight Cases...** command (or ⚖ from the toolbar) is typically used in connection with categorical data — it internally replicates rows according to the values of a **Frequency Variable**. It is useful when data is provided in the form of a cross-tabulation; see Chapter 3 for details.

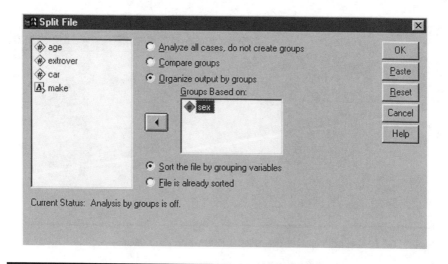

Display 1.10   **Selecting groups of cases for later analyses.**

The **Split File**... command splits rows into several groups with the effect that subsequent analyses will be carried out for each group separately. Using this command from the drop-down menu (or selecting ▦ from the toolbar) results in the dialogue box shown in Display 1.10. By default **Analyze all cases, do not create groups** is checked. A grouping of rows can be introduced by checking either **Compare groups** or **Organize output by groups** (Display 1.10). The variable (or variables) that defines the groups is specified under the **Groups Based on** list. For example, here we request that all analyses of the data shown in Display 1.2 will be carried out within gender groups of subjects. The rows of the **Data View** spreadsheet need to be sorted by the values of the grouping variable(s) for the **Split File** routine to work. It is, therefore, best to always check **Sort the file by grouping variables** on the **Split File** dialogue. Once **Split File** is activated, the status bar displays "Split File On" on the right-hand side to inform the user about this.

The **Select Cases**... command (or selecting ▦ from the toolbar) opens the dialogue shown in Display 1.11. By default **All Cases** are selected, which means that all cases in the data file are included in subsequent analyses. Checking **If condition is satisfied** allows for a subset of the cases (rows) to be selected. The condition for selection is specified using the **If...** button. This opens the **Select Cases: If** sub-dialogue box where a logical expression for evaluation can be supplied. For example, we chose to select subjects older than 40 from the gender group coded "1" (males) from the data shown in Display 1.2 which translates into using the logical expression **age > 40 & sex = 1** (Display 1.11). Once **Continue** and **OK** are pressed, the selection is activated; SPSS then "crosses out" unselected rows

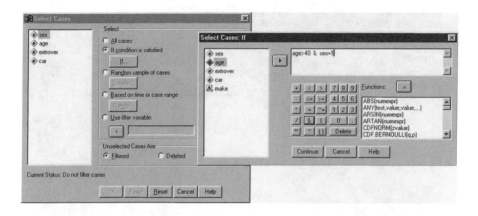

Display 1.11   Selecting subsets of cases for later analyses.

in the Data View spreadsheet and ignores these rows in subsequent analyses. It also automatically includes a filter variable, labeled filter_$ in the spreadsheet which takes the value "1" for selected rows and "0" for unselected rows. Filter variables are kept to enable replication of the case selection at a later stage by simply selecting cases for which filter_$ takes the value "1." Once the selection is active, the status bar displays "Filter On" for information. (It is also possible to remove unselected cases permanently by checking Unselected Cases Are Deleted in the Select Cases dialogue box, Display 1.11.)

1.5.2 Generating New Variables

The Transform drop-down menu provides procedures for generating new variables or changing the values of existing ones (Display 1.12).

The Compute... command is frequently used to generate variables suitable for statistical analyses or the creation of graphics. The resulting Compute dialogue can be used to create new variables or replace the values of existing ones (Display 1.13). The name of the variable to be created or for which values are to be changed is typed in the Target Variable list. For new variables, the Type&Label sub-dialogue box enables specification of variable type and label. The expression used to generate new values can be typed directly in the Expression field or constructed automatically by pasting in functions from the Functions list or selecting arithmetic operators and numbers from the "calculator list" seen in Display 1.13. When pasting in functions, the arguments indicated by question marks must be completed. Here, for example, we request a new variable, the age of a person in months (variable month), to be generated by multiplying the existing age variable in years (age) by the factor 12 (Display 1.13).

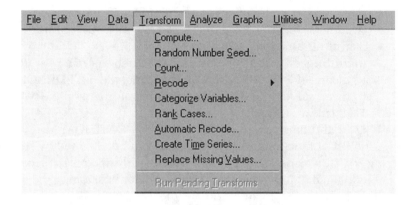

Display 1.12 Transform **drop-down menu.**

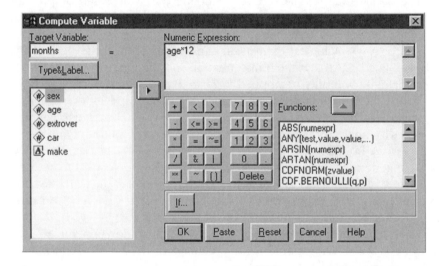

Display 1.13 Generating new variables or changing the values of existing variables.

The following applies to expressions:

■ The meaning of most arithmetic operators is obvious (+, –, *, /). Perhaps less intuitive is double star (**) for "by the power of."
■ Most of the logical operators use well-known symbols (>, = , etc.). In addition:
  ■ Ampersand (&) is used to indicate "and"
  ■ Vertical bar (|) to indicate "or"
  ■ ~= stands for "not equal"
  ■ ~ means "not" and is used in conjunction with a logical expression

- A large number of functions are supported, including
  - Arithmetic functions, such as LN(numexpr), ABS(numexpr)
  - Statistical functions, such as MEAN (numexpr, numexp,...), including distribution functions, such as CDF.NORMAL (q,mean,stddev), IDF.NORMAL (p,mean, stddev), PDF.NORMAL (q,mean,stddev); and random numbers, for example, RV.NORMAL (mean,stddev)
  - Date and time functions, for example, XDATE.JDAY(datevalue)
- A full list of functions and explanations can be obtained by searching for "functions" in the online **Help** system index. Explanations of individual functions are also provided after positioning the cursor over the function in question on the **Compute** dialogue box and right-clicking.

The **Compute Variables: If Cases** sub-dialogue box is accessed by pressing the **If...** button and works in the same way as the **Select Cases: If** sub-dialogue (see Display 1.11). It allows data transformations to be applied to selected subsets of rows. A logical expression can be provided in the field of this sub-dialogue box so that the transformation specified in the **Compute Variable** main dialogue will only be applied to rows that fulfill this condition. Rows for which the logical expression is not true are not updated.

In addition to **Compute...**, the **Recode...** command can be used to generate variables for analysis. As with the **Compute...** command, values of an existing variable can be changed (choose to recode **Into Same Variables**) or a new variable generated (choose **Into Different Variables...**). In practice, the **Recode...** command is often used to categorize continuous outcome variables and we will delay our description of this command until Chapter 3 on categorical data analysis.

The remaining commands from the **Transform** drop-down menu are used less often. We provide only a brief summary of these and exclude time series commands:

- **Random Number Seed...** allows setting the seed used by the pseudo-random number generator to a specific value so that a sequence of random numbers — for example, from a normal distribution using the function RV.NORMAL(mean,stddev) — can be replicated.
- **Count...** counts the occurrences of the same value(s) in a list of variables for each row and stores them in a new variable. This can be useful for generating summaries, for example, of repeated measures.
- **Categorize Variables...** automatically converts continuous variables into a given number of categories. Data values are categorized according to percentile groups with each group containing approximately the same number of cases.

File Edit View Data Transform	Analyze Graphs Utilities Window Help	
	Reports ▶	
	Descriptive Statistics ▶	Chapters 2, 3, 6, 9, 11
	Custom Tables ▶	
	Compare Means ▶	Chapters 2, 5
	General Linear Model ▶	Chapter 5, 6, 7
	Mixed Models ▶	Chapter 8
	Correlate ▶	Chapters 2, 4, 8
	Regression ▶	Chapters 2, 4, 9
	Loglinear ▶	
	Classify ▶	Chapter 12
	Data Reduction ▶	Chapter 11
	Scale ▶	
	Nonparametric Tests ▶	Chapters 2, 3, 5
	Time Series ▶	
	Survival ▶	Chapter 10
	Multiple Response ▶	
	Missing Value Analysis...	

Display 1.14   Statistical procedures covered in this book.

- Rank Cases... assigns ranks to variable values. Ranks can be assigned in ascending or descending order and ranking can be carried out within groups defined By a categorical variable.
- Automatic Recode... coverts string and numeric variables into consecutive integers.

1.5.3 Running Statistical Procedures

Performing a variety of statistical analyses using SPSS is the focus of this handbook and we will make extensive use of the statistical procedures offered under the **Analyze** drop-down menu in later chapters. Display 1.14 provides an overview. (There are many other statistical procedures available in SPSS that we do not cover in this book — interested readers are referred to the relevant manuals.)

1.5.4 Constructing Graphical Displays

Many (perhaps most) statistical analyses will begin by the construction of one or more graphical display(s) and so many of the commands available under the **Graphs** drop-down menu will also be used in later chapters. Display 1.15 provides an overview.

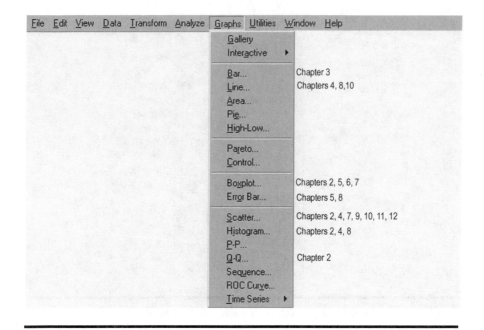

Display 1.15 **Graph procedures demonstrated in this book.**

The **Gallery** command provides a list of available charts with example displays. The **Interactive** command provides a new interactive graphing facility that we have not used in this book primarily because of space limitations. Interested readers should refer to the appropriate SPSS manuals.

1.6 The Output Viewer

Once a statistical procedure is run, an **Output Viewer** window is created that holds the results. For example, requesting simple descriptive summaries for the age and gender variables results in the output window shown in Display 1.16. Like the **Data Editor**, this window also has a menu bar and a toolbar and displays a status bar. The **File**, **Edit**, **View** and **Utilities** drop-down menus fulfill similar functions as under the **Data Editor** window, albeit with some extended features for table and chart output. The **Analyze**, **Graphs**, **Window** and **Help** drop-down menus are virtually identical. (The **Window** drop-down menu allows moving between different windows, for example between the **Output Viewer** and the **Data Editor** window in the usual way.) The **Insert** and **Format** menus provide new commands for output editing. A toolbar is provided for quick access.

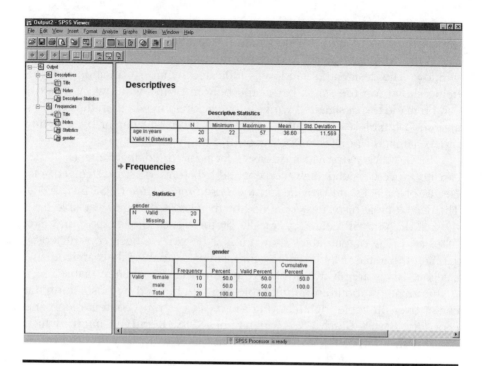

Display 1.16 Output Viewer **window.**

The **Output Viewer** is divided into two panes. The right-hand pane contains statistical tables, charts, and text output. The left-hand pane contains a tree structure similar to those used in Windows Explorer, which provides an outline view of the contents. Here, for example, we have carried out two SPSS commands, the **Descriptives** command, and the **Frequencies** command, and these define the level-1 nodes of the tree structure that then "branch out" into several output tables/titles/notes each at level 2. Level-2 displays can be hidden in the tree by clicking on the minus symbol (–) of the relevant level-1 node. Once hidden, they can be expanded again by clicking the now plus symbol (+). Clicking on an item on the tree in the left-hand pane automatically highlights the relevant part in the right-hand pane and provides a means of navigating through output.

The contents of the right-hand pane or parts of it can also be copied/pasted into other Windows applications via the **Edit** drop-down menu or the whole output saved as a file by employing the **Save** or **Save As** commands from the **File** drop-down menu. The extension used by SPSS to indicate viewer output is *.spo. (An output file can then be opened again by using **File – Open – Output**... from the menu bar.)

More than one **Output Viewer** can be open at one time. In that case SPSS directs the output into the designated **Output Viewer** window. By default this is the window opened last, rather than the active (currently selected) window. The designated window is indicated by an exclamation point (!) being shown on the status bar at the bottom of the window. A window can be made the designated window by clicking anywhere in the window and choosing **Utilities – Designate window** from the menu bar or by selecting the exclamation point symbol ![!] on the toolbar.

The **Output Viewer** provides extensive facilities for editing contents. Tables can be moved around, new contents added, fonts or sizes changed, etc. Details of facilities are provided in the *Base User's Guide* (SPSS Inc., 2001). The default table display is controlled by the **Options** dialogue available from the **Edit** drop-down menu, specifically by the **Viewer, Output Labels,** and **Pivot Tables** tabs. The output tables shown in this book have been constructed by keeping the initial default settings, for example, variable labels and variable category labels are always displayed in output tables when available.

Information about an output table can be obtained by positioning the cursor over the table, right-clicking to access a pop-up context menu and choosing **Results Coach**. This opens the SPSS **Results Coach,** which explains the purpose of the table and the contents of its cells, and offers information on related issues.

Whenever an analysis command is executed, SPSS produces a "Notes" table in the **Output Viewer**. By default this table is hidden in the right-hand pane display, but any part of the output can be switched between background (hidden) and foreground (visible) by double-clicking on its book icon in the tree structure in the left-hand pane. The "Notes" table provides information on the analysis carried out — data file used, analysis command syntax, time of execution, etc. are all recorded.

Most output displayed in tables (in so-called pivot tables) can be modified by double-clicking on the table. This opens the **Pivot Table Editor,** which provides an advanced facility for table editing (for more details see the *Base User's Guide*, SPSS Inc., 2001d). For example, table cell entries can be edited by double-clicking onto the respective cell or columns collapsed by dragging their borders. Display options can be accessed from the editor's own menu bar or from a context menu activated by right-clicking anywhere within the **Pivot Table Editor.** One option in the context menu automatically creates a chart. While this might appear convenient at first, it rarely produces an appropriate graph. A more useful option is to select **What's this** from the pop-up context menu for cells with table headings that will provide an explanation of the relevant table entries.

Text output not displayed in pivot tables can also be edited to some extent. Double-clicking on the output opens the **Text Output Editor,** which allows for editing the text and changing the font characteristics.

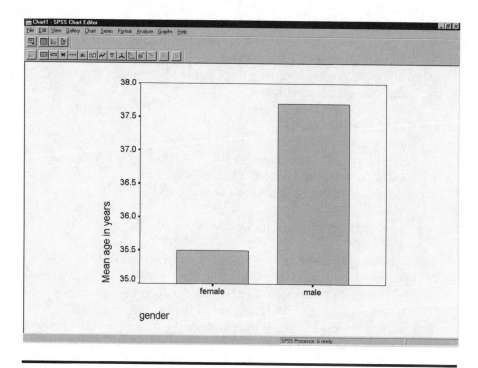

Display 1.17   Chart Editor **window.**

1.7 The Chart Editor

The use of procedures from the **Graphs** drop-down menu and some procedures from the **Analyze** menu generate chart output in the **Output Viewer.** After creating a chart, it is often necessary to modify it, for example, to enhance it for presentation or to obtain additional information. In SPSS this can be done by activating the **Chart Editor** by double-clicking on the initial graph in the **Output Viewer.** As an example, a bar chart of mean ages within gender has been created for the data in Display 1.2 and is displayed in the **Chart Editor** (Display 1.17).

The **Chart Editor** has its own menu bar and toolbar. (However, the **Analyze, Graphs,** and **Help** drop-down menus remain unchanged.) Once a graph is opened in the **Chart Editor,** it can be edited simply by double-clicking on the part that is to be changed, for example, an axis label. Double-clicking opens dialogue boxes that can alternatively be accessed via the menus. Specifically, the **Gallery** drop-down menu provides a facility for converting between different types of graphs; the **Chart** drop-down menu deals mainly with axes, chart and text displays; and the **Format** drop-down with colors, symbols, line styles, patterns, text fonts, and sizes.

The **Chart Editor** facilities are described in detail in the *Base User's Guide* (SPSS Inc., 2001). Here we provide only an introductory editing example. In later chapters we will explain more facilities as the need arises.

As an example we attempt to enhance the bar chart in Display 1.17. In particular, this graphical display:

1. Should not be in a box
2. Should have a title
3. Should have different axes titles
4. Should be converted into black and white
5. Should have a *y*-axis that starts at the origin

Making these changes requires the following steps:

1. Uncheck **Inner Frame** on the **Chart** drop-down menu.
2. Use the command **Title...** from the **Chart** drop-down menu. This opens the **Titles** dialogue box where we type in our chosen title "Bar chart" and set **Title Justification** to **Center.**
3. Double-click on the *y*-axis title, change the **Axis Title** in the resulting **Scale Axis** dialogue box, and also set **Title Justification** to **Center** in that box. We then double-click on the *x*-axis label and again change the **Axis Title** in the resulting **Category Axis** dialogue box and also set **Title Justification** to **Center** in that box.
4. Select the bars (by single left-click), choose **Color...** from the **Format** drop-down menu and select white fill in the resulting **Colors** palette. We also choose **Fill Pattern...** from the **Format** drop-down menu and apply a striped pattern from the resulting **Fill Patterns** palette.
5. Double-click on the *y*-axis and change the **Minimum Range** to "0" and the **Maximum Range** to "60" in the resulting **Scale Axis** dialogue box. With this increased range, we also opt to display **Major Divisions** at an increment of "10." Finally, we employ the **Labels** sub-dialogue box to change the Decimal Places to "0."

The final bar chart is shown in Display 1.18.

Graphs can be copied and pasted into other applications via the **Edit** drop-down menu from the **Chart Editor** (this is what we have done in this book) or from the **Data Viewer**. Graphs can also be saved in a number of formats by using **File – Export Chart** from the **Chart Editor** menu bar. The possible formats are listed under the **Save as type** list of the **Export Chart** dialogue box. In SPSS version 11.0.1 they include JPEG (*.jpg), PostScript (*.eps), Tagged Image File (*.tif), and Windows Metafile (*.wmf).

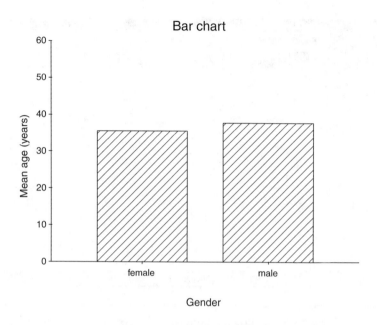

Display 1.18 Bar chart.

1.8 Programming in SPSS

Most commands are accessible from the menus and dialogue boxes. However, some commands and options are available only by using SPSS's command language. It is beyond the scope of this book to cover the command syntax; we refer the reader to the *Syntax Reference Guide* (SPSS, Inc., 2001c) for this purpose.

It is useful, however, to show how to generate, save, and run command syntax. From an organizational point of view, it is a good idea to keep a record of commands carried out during the analysis process. Such a record ("SPSS program") allows for quick and error-free repetition of the analysis at a later stage, for example, to check analysis results or update them in line with changes to the data file. This also allows for editing the command syntax to utilize special features of SPSS not available through dialogue boxes.

Without knowing the SPSS command syntax, a syntax file can be generated by employing the **Paste** facility provided by all dialogue boxes used for aspects of statistical analysis. For example, the main dialogue boxes shown in Displays 1.8, 1.10, 1.11, and 1.13 all have a **Paste** command button. Selecting this button translates the contents of a dialogue box and related sub-dialogue boxes into command syntax. The command with

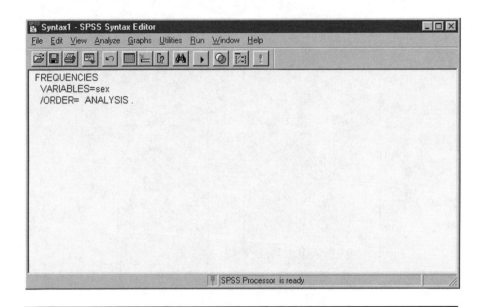

Display 1.19 Syntax Editor **showing command syntax for the dialogue box in Display 1.8.**

options is pasted into a Syntax Editor window. Should such a window not exist at the time, SPSS will automatically create one. For example, selecting Paste on the dialogue box shown in Display 1.8 produces the Syntax Editor window given in Display 1.19. If a Syntax Editor window already exists, then SPSS will append the latest command to the contents of the window.

The commands contained in a Syntax Editor window can be executed by selecting the command All... from the Run drop-down menu on the editor's menu bar or by selecting ▶ from the toolbar. This generates output in the Output Viewer in the usual way. It is also possible to execute only selected commands in the syntax window by highlighting the relevant commands and then using the command Selection from the Run drop-down menu or by clicking the run symbol on the toolbar.

The contents of the Syntax Editor window can be saved as a syntax file by using the Save or Save As command from the File drop-down menu of the Syntax Editor. The extension used by SPSS to indicate syntax is *.sps. A syntax file can be opened again by using the commands File – Open – Syntax... from the Data Editor, Output Viewer, or Syntax Editor menu bar.

More than one Syntax Editor can be open at one time. In that case, SPSS executes the commands of the designated Syntax Editor window. By default, the window opened last is the designated window. The designation is indicated and can be changed in the same way as that of the Output Viewer windows (see Section 1.6).

Chapter 2

Data Description and Simple Inference for Continuous Data: The Lifespans of Rats and Ages at Marriage in the U.S.

2.1 Description of Data

In this chapter, we consider two data sets. The first, shown in Table 2.1, involves the lifespan of two groups of rats, one group given a restricted diet and the other an *ad libitum* diet (that is, "free eating"). Interest lies in assessing whether lifespan is affected by diet.

The second data set, shown in Table 2.2, gives the ages at marriage for a sample of 100 couples that applied for marriage licences in Cumberland County, PA, in 1993. Some of the questions of interest about these data are as follows:

- How is age at marriage distributed?
- Is there a difference in average age at marriage of men and women?
- How are the ages at marriage of husband and wife related?

Table 2.1   Lifespans of Rats (in Days) Given Two Diets

a) Restricted diet (*n* = 105)

105	193	211	236	302	363	389	390	391	403	530	604	60.5	630	716
718	727	731	749	769	770	789	804	810	811	833	868	871	848	893
897	901	906	907	919	923	931	940	957	958	961	962	974	979	982
1101	1008	1010	1011	1012	1014	1017	1032	1039	1045	1046	1047	1057	1063	1070
1073	1076	1085	1090	1094	1099	1107	1119	1120	1128	1129	1131	1133	1136	1138
1144	1149	1160	1166	1170	1173	1181	1183	1188	1190	1203	1206	1209	1218	1220
1221	1228	1230	1231	1233	1239	1244	1258	1268	1294	1316	1327	1328	1369	1393
1435														

b) *Ad libitum* diet (*n* = 89)

89	104	387	465	479	494	496	514	532	536	545	547	548	582	606
609	619	620	621	630	635	639	648	652	653	654	660	665	667	668
670	675	677	678	678	681	684	688	694	695	697	698	702	704	710
711	712	715	716	717	720	721	730	731	732	733	735	736	738	739
741	743	746	749	751	753	764	765	768	770	773	777	779	780	788
791	794	796	799	801	806	807	815	836	838	850	859	894	963	

Source: Berger, Boss, and Guess, 1988. With permission of the Biometrics Society.

2.2 Methods of Analysis

Data analysis generally begins with the calculation of a number of summary statistics such as the *mean, median, standard deviation*, etc., and by creating informative graphical displays of the data such as *histograms, box plots,* and *stem-and-leaf plots.* The aim at this stage is to describe the general distributional properties of the data, to identify any unusual observations (*outliers*) or any unusual patterns of observations that may cause problems for later analyses to be carried out on the data. (Descriptions of all the terms in italics can be found in Altman, 1991.)

Following the initial exploration of the data, statistical tests may be applied to answer specific questions or to test particular hypotheses about the data. For the rat data, for example, we will use an *independent samples t-test* and its nonparametric alternative, the *Mann-Whitney U-test* to assess whether the average lifetimes for the rats on the two diets differ. For the second data set we shall apply a *paired samples t-test* (and the *Wilcoxon signed ranks test*) to address the question of whether men and women have different average ages at marriage. (See Boxes 2.1 and 2.2 for a brief account of the methods mentioned.)

Finally, we shall examine the relationship between the ages of husbands and their wives by constructing a *scatterplot*, calculating a number of *correlation coefficients*, and fitting a simple *linear regression model* (see Box 2.3).

Table 2.2   Ages (in years) of Husbands and Wives at Marriage

Husband	Wife	Husband	Wife	Husband	Wife	Husband	Wife	Husband	Wife
22	21	40	46	23	22	31	33	24	25
38	42	26	25	51	47	23	21	25	24
31	35	29	27	38	33	25	25	46	37
42	24	32	39	30	27	27	25	24	23
23	21	36	35	36	27	24	24	18	20
55	53	68	52	50	55	62	60	26	27
24	23	19	16	24	21	35	22	25	22
41	40	52	39	27	34	26	27	29	24
26	24	24	22	22	20	24	23	34	39
24	23	22	23	29	28	37	36	26	18
19	19	29	30	36	34	22	20	51	50
42	38	54	44	22	26	24	27	21	20
34	32	35	36	32	32	27	21	23	23
31	36	22	21	51	39	23	22	26	24
45	38	44	44	28	24	31	30	20	22
33	27	33	37	66	53	32	37	25	32
54	47	21	20	20	21	23	21	32	31
20	18	31	23	29	26	41	34	48	43
43	39	21	22	25	20	71	73	54	47
24	23	35	42	54	51	26	33	60	45

Source: Rossman, 1996. With permission of Springer-Verlag.

Box 2.1 Student's t-Tests

(1) *Independent samples t-test*

■ The independent samples t-test is used to test the null hypothesis that the means of two populations are the same, $H_0: \mu_1 = \mu_2$, when a sample of observations from each population is available. The observations made on the sample members must all be independent of each other. So, for example, individuals from one population must not be individually matched with those from the other population, nor should the individuals within each group be related to each other.

■ The variable to be compared is assumed to have a normal distribution with the same standard deviation in both populations.

■ The test-statistic is

$$t = \frac{\bar{y}_1 - \bar{y}_2}{s\sqrt{\dfrac{1}{n_1} + \dfrac{1}{n_2}}}$$

where $\bar{y}_1$ and $\bar{y}_2$ are the means in groups 1 and 2, n_1 and n_2 are the sample sizes, and s is the pooled standard deviation calculated as

$$s = \sqrt{\frac{(n_1 - 1)s_1^2 + (n_2 - 1)s_2^2}{n_1 + n_2 - 2}}$$

where s_1 and s_2 are the standard deviations in the two groups.
■ Under the null hypothesis, the t-statistic has a student's t-distribution with $n_1 + n_2 - 2$ degrees of freedom.
■ The confidence interval corresponding to testing at the α significance level, for example, if $\alpha = 0.05$, a 95% confidence interval is constructed as

$$(\bar{y}_1 - \bar{y}_2) \pm t_\alpha s \sqrt{\frac{1}{n_1} + \frac{1}{n_2}}$$

when t_α is the critical value for a two-sided test, with $n_1 + n_2 - 2$ degrees of freedom.

(2) *Paired samples t-test*

■ A paired t-test is used to compare the means of two populations when samples from the populations are available, in which each individual in one sample is paired with an individual in the other sample. Possible examples are anorexic girls and their healthy sisters, or the same patients before and after treatment.
■ If the values of the variables of interest y for the members of the ith pair in groups 1 and 2 are denoted as y_{1i} and y_{2i}, then the differences $d_i = y_{1i} - y_{2i}$ are assumed to have a normal distribution.

■ The null hypothesis here is that the mean difference is zero, i.e., $H_0{:}\mu_d = 0$.
■ The paired t-statistic is

$$t = \frac{\bar{d}}{s_d/\sqrt{n}}$$

where $\bar{d}$ is the mean difference between the paired groups and s_d is the standard deviation of the differences d_i and n the number of pairs. Under the null hypothesis, the test-statistic has a t-distribution with $n - 1$ degrees of freedom.
■ A $100(1 - \alpha)\%$ confidence interval can be constructed as follows:

$$\bar{d} \pm t_\alpha s_d/\sqrt{n}$$

where t_α is the critical value for a two-sided test with $n - 1$ degrees of freedom.

Box 2.2 Nonparametric Tests

(1) *Mann-Whitney U-test*

■ The null hypothesis to be tested is that the two populations being compared have identical distributions. (For two normally distributed populations with common variance, this would be equivalent to the hypothesis that the means of the two populations are the same.)
■ The alternative hypothesis is that the population distributions differ in location (the median).
■ Samples of observations are available from each of the two populations being compared.
■ The test is based on the joint ranking of the observations from the two samples (as if they were from a single sample). If there are ties, the tied observations are given the average of the ranks for which the observations are competing.
■ The test statistic is the sum of the ranks of one sample (the lower of the two rank sums is generally used).

■ For small samples, p-values for the test-statistic can be found from suitable tables (see Hollander and Wolfe, 1999).

■ A large sample approximation is available that is suitable when the two sample sizes n_1 and n_2 are both greater than 15, and there are no ties. The test-statistic z is given by

$$z = \frac{S - n_1(n_1 + n_2 + 1)/2}{\sqrt{n_1 n_2(n_1 + n_2 + 1)/12}}$$

where S is the test-statistic based on the sample with n_1 observations. Under the null hypothesis, z has approximately a standard normal distribution.

■ A modified z-statistic is available when there are ties (see Hollander and Wolfe, 1999).

(2) *Wilcoxon signed ranks test*

■ Assume, we have two observations, y_{i1} and y_{i2} on each of n subjects in our sample, e.g., before and after treatment. We first calculate the differences $z_i = y_{i1} - y_{i2}$, between each pair of observations.

■ To compute the Wilcoxon signed-rank statistic T^+, form the absolute values of the differences z_i and then order them from least to greatest.

■ If there are ties among the calculated differences, assign each of the observations in a tied group the average of the integer ranks that are associated with the tied group.

■ Now assign a positive or negative sign to the ranks of the differences according to whether the corresponding difference was positive or negative. (Zero values are discarded, and the sample size n altered accordingly.)

■ The statistic T^+ is the sum of the positive ranks. Tables are available for assigning p-values (see Table A.4 in Hollander and Wolfe, 1999).

■ A large sample approximation involves testing the statistic z as a standard normal:

$$z = \frac{T^+ - n(n + 1)/4}{\sqrt{n(n + 1)(2n + 1)/24}}$$

Box 2.3 Simple Linear Regression

- Simple linear regression is used to model the relationship between a single response variable, y, and a single explanatory variable, x; the model is

$$y_i = \beta_0 + \beta_1 x_i + \varepsilon_i$$

where (x_i, y_i), $i = 1, \ldots, n$ are the sample values of the response and exploratory variables and ε_i are random disturbance terms assumed to be normally distributed with mean zero and variance σ^2.

- The intercept parameter, β_0, is the value predicted for the response variable when the explanatory variable takes the value zero.
- The slope parameter, β_1, is the change in the response variable predicted when the explanatory variable is increased by one unit.
- The parameters, also known as *regression coefficients*, can be estimated by least squares (see Rawlings, Pantula, and Dickey, 1998).

2.3 Analysis Using SPSS

2.3.1 Lifespans of Rats

The lifespan data from Table 2.1 can be typed directly into the SPSS **Data View** as shown in Display 2.1. SPSS assumes that the rows of the spreadsheet define the units on which measurements are taken. Thus the rats have to correspond to rows and the lifespan values have to be included in one long column. For later identification, a number is also assigned to each rat.

The type of diet given to each rat has to be designated as a *factor*, i.e., a categorical variable with, in this case, two levels, "Restricted diet" and "*Ad libitum* diet." As mentioned before (cf. Chapter 1), factor levels need to be assigned artificial codes and labels attached to level codes rather than text typed directly into the **Data View** spreadsheet. We therefore use the codes "1" and "2" in the **Data View** spreadsheet (Display 2.1) and employ the **Values** column of the **Variable View** spreadsheet to attach labels "Restricted diet" and "Ad libitum diet" to the codes (see Display 1.4).

	rat	lifespan	diet
89	89	1218	1
90	90	1220	1
91	91	1221	1
92	92	1228	1
93	93	1230	1
94	94	1231	1
95	95	1233	1
96	96	1239	1
97	97	1244	1
98	98	1258	1
99	99	1268	1
100	100	1294	1
101	101	1316	1
102	102	1327	1
103	103	1328	1
104	104	1369	1
105	105	1393	1
106	106	1435	1
107	107	89	2
108	108	104	2
109	109	387	2
110	110	465	2
111	111	479	2
112	112	494	2
113	113	496	2
114	114	514	2
115	115	532	2
116	116	536	2
117	117	545	2
118	118	547	2

Display 2.1   Data View **spreadsheet for rat data from Table 2.1.**

The analysis of almost every data set should begin by examination of relevant summary statistics, and a variety of graphical displays. SPSS supplies standard summary measures of location and spread of the distribution of a continuous variable together with a variety of useful graphical displays. The easiest way to obtain the data summaries is to select the commands

Analyze – Descriptive statistics – Explore

from the menu bar and to fill in resulting dialogue boxes as indicated in Display 2.2. In this box, the Dependent List declares the continuous variables

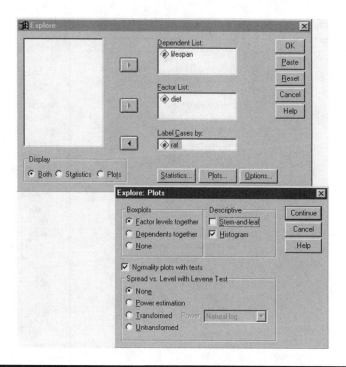

Display 2.2   Generating descriptive statistics within groups.

requiring descriptive statistics, and in the **Factor List** we specify the categories within which separate summaries are to be calculated — in this example, each dietary group. Labeling the observations by the rat's ID number will enable possible outlying observations to be identified.

For graphical displays of the data we again need the **Explore** dialogue box; in fact, by checking **Both** in this box, we can get our descriptive statistics *and* the plots we require. Here we select **Boxplots** and **Histogram** to display the distributions of the lifespans of the rats, and *probability plots* (see Everitt and Wykes, 1999) to assess more directly the assumption of normality within each dietary group.

Display 2.3 shows the descriptive statistics supplied by default (further statistics can be requested from **Explore** via the **Statistics** sub-dialogue box). We see, for example, that the median lifespan is shorter for rats on the *ad libitum* diet (710 days compared with 1035.5 days for rats on the restricted diet). A similar conclusion is reached when either the mean or the *5% trimmed mean* (see Everitt and Wykes, 1999) is used as the measure of location. The "spread" of the lifespans as measured by the *interquartile range* (IQR) (see Everitt and Wykes, 1999) appears to vary with diet, with lifespans in the restricted diet group being more variable (IQR in the restricted diet group is 311.5 days, but only 121 days in the comparison

Descriptives

diet				Statistic	Std. Error
lifespan in days	Restricted diet	Mean		968.75	27.641
		95% Confidence Interval for Mean	Lower Bound	913.94	
			Upper Bound	1023.55	
		5% Trimmed Mean		988.31	
		Median		1035.50	
		Variance		80985.696	
		Std. Deviation		284.580	
		Minimum		105	
		Maximum		1435	
		Range		1330	
		Interquartile Range		311.50	
		Skewness		-1.161	.235
		Kurtosis		1.021	.465
	Ad libitum diet	Mean		684.01	14.213
		95% Confidence Interval for Mean	Lower Bound	655.77	
			Upper Bound	712.26	
		5% Trimmed Mean		695.05	
		Median		710.00	
		Variance		17978.579	
		Std. Deviation		134.084	
		Minimum		89	
		Maximum		963	
		Range		874	
		Interquartile Range		121.00	
		Skewness		-2.010	.255
		Kurtosis		7.027	.506

Display 2.3   Descriptives output for rat data.

group). Other measures of spread, such as the standard deviation and the range of the sample, confirm the increased variability in the restricted diet group.

Finally, SPSS provides measures of two aspects of the "shape" of the lifespan distributions in each dietary group, namely, *skewness* and *kurtosis* (see Everitt and Wykes, 1999). The index of skewness takes the value zero for a *symmetrical distribution*. A negative value indicates a *negatively skewed distribution*, a positive value a *positively skewed distribution* — Figure 2.1 shows an example of each type. The kurtosis index measures the extent to which the peak of a unimodal frequency distribution departs from the shape of normal distribution. A value of zero corresponds to a normal distribution; positive values indicate a distribution that is more pointed than a normal distribution and a negative value a flatter distribution — Figure 2.2 shows examples of each type. For our data we find that the two shape indices indicate some degree of negative skewness

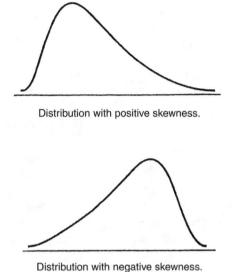

Distribution with positive skewness.

Distribution with negative skewness.

Figure 2.1   Examples of skewed distributions.

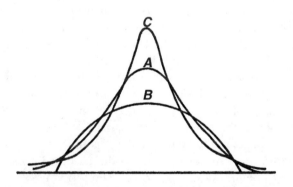

Figure 2.2   Curves with different degrees of kurtosis.

and distributions that are more pointed than a normal distribution. Such findings have possible implications for later analyses that may be carried out on the data.

We can now move on to examine the graphical displays we have selected. The box plots are shown in Display 2.4. (We have edited the

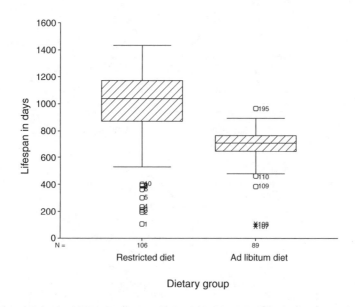

Display 2.4   Box plots for rat lifespans generated by commands in Display 2.2.

original graph somewhat using the Chart Editor to improve its appearance.) This type of plot (also known as *box-and-whisker plot*) provides a "picture" of a five-point summary of the sample observations in each group. The lower end of the box represents the lower quartile and the upper end the upper quartile; thus the box width is the IQR and covers the middle 50% of the data. The horizontal line within the box is placed at the median of the sample. The bottom "whisker" extends to the minimum data point in the sample, except if this point is deemed an outlier by SPSS. (SPSS calls a point an "outlier" if the point is more than 1.5 × IQR away from the box and considers it an "extreme value" when it is more than 3 × IQR away.) In the latter case, the whisker extends to the second lowest case, except if this is found to be an outlier and so on. The top whisker extends to the maximum value in the sample, again provided this value is not an outlier.

The box plots in Display 2.4 lead to the same conclusions as the descriptive summaries. Lifespans in the restricted diet group appear to be longer "on average" but also more variable. A number of rats have been indicated as possible outliers and for the *ad libitum* diet; some are even marked as extreme observations. Since we have employed case labels, we can identify the rats with very short lifespans. Here the rat with the shortest lifespan (89 days) is rat number 107. (Lifespans that are short relative to the bulk of the data can arise as a result of negative skewness of the distributions — observations labeled "outliers" by SPSS do not

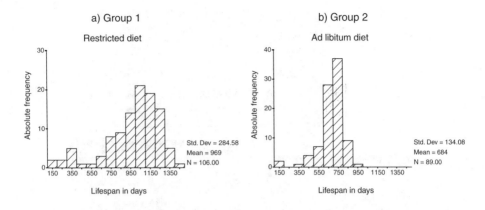

Display 2.5   Histograms for rat lifespans generated by commands in Display 2.2.

necessarily have to be removed before further analyses, although they do merit careful consideration. Here we shall *not* remove any of the suspect observations before further analyses.)

The evidence from both the summary statistics for the observations in each dietary group and the box plot is that the distributions of the lifespans in the underlying population are nonsymmetric and that the variances of the lifespans vary between the diet groups. Such information is important in deciding which statistical tests are most appropriate for testing hypotheses of interest about the data, as we shall see later.

An alternative to the box plot for displaying sample distributions is the *histogram*. Display 2.5 shows the histograms for the lifespans under each diet. Each histogram displays the frequencies with which certain ranges (or "bins") of lifespans occur within the sample. SPSS chooses the bin width automatically, but here we chose both our own bin width (100 days) and the range of the *x*-axis (100 days to 1500 days) so that the histograms in the two groups were comparable. To change the default settings to reflect our choices, we go through the following steps:

- Open the **Chart Editor** by double clicking on the graph in the **Output Viewer.**
- Double click on the *x*-axis labels.
- Check **Custom** and select **Define...** in the resulting **Interval Axis** box (Display 2.6).
- Alter the **Interval Axis: Define Custom In...** sub-dialogue box as shown in Display 2.6 and click **Continue.**
- Select **Labels...** in the **Interval Axis** box.
- Change the settings of the **Interval Axis: Labels** sub-dialogue box as shown in Display 2.6 and finally click **Continue** followed by **OK.**

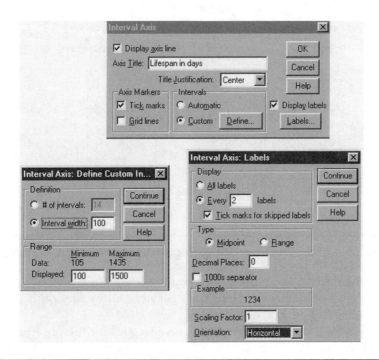

Display 2.6   Settings for controlling the bin width in the histograms shown in Display 2.5.

As we might expect the histograms indicate negatively skewed frequency distributions with the left-hand tail being more pronounced in the restricted diet group.

Finally, normality can be assessed more formally with the help of a *quantile-quantile probability plot* (Q-Q plot); this involves a plot of the quantiles expected from a standard normal distribution against the observed quantiles (for more details see Everitt, 2001b). Such a plot for the observations in each group is shown in Display 2.7. A graph in which the points lie approximately on the reference line indicates normality. Points above the line indicate that the observed quantiles are lower than expected and vice versa. For the rats lifespan data we find that the very small and very large quantiles are smaller than would be expected — with this being most pronounced for the lowest three quantiles in the *ad libitum* group. Such a picture is characteristic of distributions with a heavy left tail; thus again we detect some degree of negative skewness.

Having completed an initial examination of the lifespan data, we now proceed to assess more formally whether the average lifespan is the same under each dietary regime. We can do this using an appropriate significance

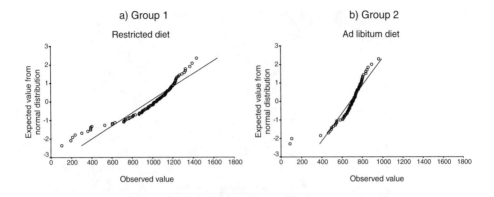

Display 2.7   Q-Q plots for rat lifespans generated by commands in Display 2.2.

test and by constructing the relevant confidence interval. To begin, we shall apply an independent samples Student's *t*-test (see Box 2.1), conveniently ignoring for the moment the indication given by our preliminary examination of the data that two of the assumptions of the test, normality and homogeneity of variance, might not be strictly valid.

The independent samples *t*-test assesses the null hypothesis that the population means of lifespan in each of the two diet groups are equal (see Box 2.1). The test can be accessed in SPSS by selecting the commands

Analyze – Compare Means – Independent-Samples T Test

from the menu bar to give the dialogue box shown in Display 2.8. The **Test Variable(s)** list contains the variables that are to be compared between two levels of the **Grouping Variable**. Here the variable **lifespan** is to be compared between level "1" and "2" of the grouping variable **diet**. The **Define Groups...** sub-dialogue box is used to define the levels of interest. In this example, the grouping variable has only two levels, but pair-wise group comparisons are also possible for variables with more than two group levels.

Display 2.9 shows the resulting SPSS output. This begins with a number of descriptive statistics for each group. (Note that in contrast to the output in Display 2.3 the standard errors of the means are given, i.e., the standard deviation of lifespan divided by the square root of the group sample size.) The next part of the display gives the results of applying *two* versions of the independent samples *t*-test; the first is the usual form, based on assuming equal variances in the two groups (i.e., homogeneity of variance),

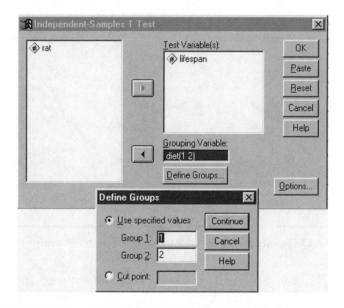

Display 2.8   Generating an independent samples *t*-test.

Group Statistics

	diet	N	Mean	Std. Deviation	Std. Error Mean
lifespan in days	Restricted diet	106	968.75	284.580	27.641
	Ad libitum diet	89	684.01	134.084	14.213

Independent Samples Test

		Levene's Test for Equality of Variances		t-test for Equality of Means						
									95% Confidence Interval of the Difference	
		F	Sig.	t	df	Sig.(2-tailed)	Mean Difference	Std. Error Difference	Lower	Upper
lifespan in days	Equal variances assumed	33.433	.000	8.664	193	.000	284.73	32.866	219.912	349.556
	Equal variances not assumed			9.161	154.940	.000	284.73	31.081	223.337	346.131

Display 2.9   Independent samples *t*-test output.

the second allows the variances to be different (this version of the test is described in detail in Everitt and Rabe-Hesketh, 2001).

The results given in the first row ($t = 8.7$, df $= 193$, $p < 0.001$), under the column heading "t-test for Equality of Means" are those for the standard *t*-test and assume homogeneity of variance. The conclusion from the test is that there is strong evidence of a difference in the mean lifespan under the two dietary regimes. Further output gives an estimate for the mean difference in lifespans between the two diets (284.7 days), and uses the standard error of this estimator (32.9 days) to construct a 95% CI for the mean difference (from 219.9 to 349.6 days). The mean lifespan in the restricted diet group is between about 220 and 350 days longer than the corresponding value in the *ad libitum* diet.

The "Independent Samples Test" table also includes a statistical significance test proposed by Levene (1960) for testing the null hypothesis that the variances in the two groups are equal. In this instance, the test suggests that there is a significant difference in the size of the within diet variances ($p < 0.001$). Consequently, it may be more appropriate here to use the alternative version of the *t*-test given in the second row of the table. This version of the *t*-test uses separate variances instead of a pooled variance to construct the standard error and reduces the degrees of freedom to account for the extra variance estimated (for full details, see Everitt and Rabe-Hesketh, 2001). The conclusion from using the second type of *t*-test is almost identical in this case.

Since earlier analyses showed there was some evidence of abnormality in the lifespans data, it may be useful to look at the results of an appropriate nonparametric test that does not rely on this assumption. SPSS supplies the Mann-Whitney U-test (also known as the *Wilcoxon rank sum test*), a test for comparing the locations of the distributions of two groups based on the ranks of the observations (see Box 2.2 for details). Such nonparametric tests can be accessed in SPSS using

Analyze – Nonparametric Tests

from the menu bar. Choosing **Two Independent Samples** then supplies the Dialogue box shown in Display 2.10. The settings are analoguous to those for the independent samples *t*-test.

The resulting output is shown in Display 2.11. All observations are ranked as if they were from a single sample and the Wilcoxon W test statistic is the sum of the ranks in the smaller group. SPSS displays the sum of the ranks together with the mean rank so that the values can be compared between groups. By default, the software uses the normal approximation (but see also Exercise 2.4.5). The Mann-Whitney test ($z = 8.3$, $p < 0.001$) also indicates a significant group difference consistent with the earlier results from the *t*-test.

2.3.2 Husbands and Wives

The data in Table 2.2 can be typed directly into a **Data View** spreadsheet as shown in Display 2.12. In this data set, it is essential to appreciate the "paired" nature of the ages arising from each couple. In order for SPSS to deal with this paired structure appropriately, the ages at marriage of a

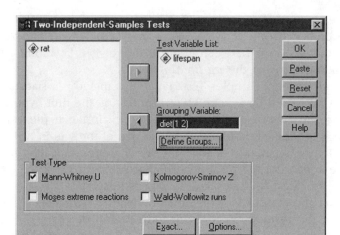

Display 2.10 Generating a Mann-Whitney U-test.

Ranks

	diet	N	Mean Rank	Sum of Ranks
lifespan in days	Restricted diet	106	128.85	13658.50
	Ad libitum diet	89	61.25	5451.50
	Total	195		

Test Statistics [a]

	lifespan in days
Mann-Whitney U	1446.500
Wilcoxon W	5451.500
Z	-8.332
Asymp. Sig. (2-tailed)	.000

[a.] Grouping Variable: diet

Display 2.11 Output generated by commands in Display 2.10.

husband and his wife must appear in the same row. For identification purposes, it is useful to include a couple number.

As in the previous example, the first step in the analysis of these data will be to calculate summary statistics for each of the two variables, age at marriage of husbands and age at marriage of wives. This example differs from the previous one in that the observational units, i.e., the married couples, do not fall into two groups. Rather, we have two age outcomes. Therefore, this time we select the commands

	couple	husband	wife
1	1.00	22.00	21.00
2	2.00	38.00	42.00
3	3.00	31.00	35.00
4	4.00	42.00	24.00
5	5.00	23.00	21.00
6	6.00	55.00	53.00
7	7.00	24.00	23.00
8	8.00	41.00	40.00
9	9.00	26.00	24.00
10	10.00	24.00	23.00
11	11.00	19.00	19.00
12	12.00	42.00	38.00
13	13.00	34.00	32.00
14	14.00	31.00	36.00
15	15.00	45.00	38.00
16	16.00	33.00	27.00
17	17.00	54.00	47.00
18	18.00	20.00	18.00
19	19.00	43.00	39.00
20	20.00	24.00	23.00
21	21.00	40.00	46.00
22	22.00	26.00	25.00
23	23.00	29.00	27.00
24	24.00	32.00	39.00
25	25.00	36.00	35.00
26	26.00	68.00	52.00
27	27.00	19.00	16.00
28	28.00	52.00	39.00
29	29.00	24.00	22.00

Display 2.12 Data View **spreadsheet for ages at marriage data from Table 2.2.**

Analyze – Descriptive Statistics – Frequencies…

from the menu bar to allow us to calculate some summary measures for both age variables (Display 2.13).

Display 2.14 shows the resulting output. The median age at marriage of husbands is 29 years, that of wives somewhat lower at 27 years. There is considerable variability in both the ages of husbands and wives, ranging as they do from late teens to early 70s.

The first question we want to address about these data is whether there is any evidence of a systematic difference in the ages of husbands and wives. Overall, wives are younger at marriage in our sample. Could this observed difference have occurred by chance when the typical age difference between husbands and wives in Cumberland County is zero?

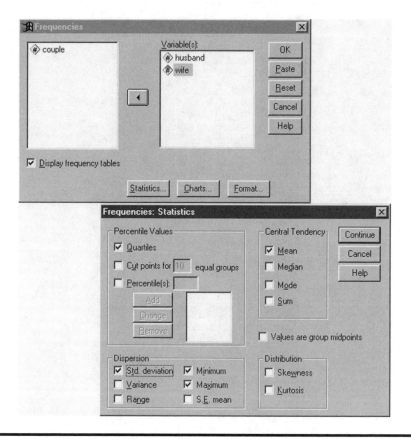

Display 2.13   Generating descriptive statistics for several variables.

Statistics

		husbands' ages at marriage	wives' ages at marriage
N	Valid	100	100
	Missing	0	0
Mean		33.0800	31.1600
Std. Deviation		12.31053	11.00479
Minimum		18.00	16.00
Maximum		71.00	73.00
Percentiles	25	24.0000	22.2500
	50	29.0000	27.0000
	75	39.5000	38.0000

Display 2.14   Descriptive output for ages at marriage.

```
           DIFFEREN Stem-and-Leaf Plot

        Frequency    Stem &  Leaf

            5.00       -7 .  00000
            1.00       -6 .  0
            4.00       -5 .  0000
            4.00       -4 .  0000
            1.00       -3 .  0
            4.00       -2 .  0000
            8.00       -1 .  00000000
             .00       -0 .
            6.00        0 .  000000
           20.00        1 .  00000000000000000000
           15.00        2 .  000000000000000
            6.00        3 .  000000
            4.00        4 .  0000
            4.00        5 .  0000
            2.00        6 .  00
            4.00        7 .  0000
            2.00        8 .  00
            2.00        9 .  00
            1.00       10 .  0
            7.00 Extremes    (>=12.0)

        Stem width:      1.00
        Each leaf:       1 case(s)
```

Display 2.15   Stem-and-leaf plot for age differences.

A possibility for answering these questions is the paired samples *t*-test (see Box 2.1). This is essentially a test of whether, in the population, the differences in age of the husbands and wives in married couples have a zero mean. The test assumes that the population differences have a normal distribution. To assess this assumption, we first need to calculate the age difference of each couple in our sample. We include a new variable, differen, in our **Data View** spreadsheet by using the **Compute** command with the formula differen = husband-wife (see Chapter 1, Display 1.13).

The observed differences can now be examined in a number of ways, including using the box plot or histogram as described earlier. But to avoid repetition here, we shall use a stem-and-leaf plot. SPSS supplies the plot from the **Explore** box as shown in Display 2.2. In this case, the **Dependent List** simply contains the variable difference, the **Factor List** is left empty and the **Stem-and-leaf** option is chosen in the **Plots** sub-dialogue box. The resulting plot is given in Display 2.15.

The stem-and-leaf plot is similar to a histogram rotated by 90°. However, in addition to the shape of the distribution, it also lists the actual values taken by each observation, by separating each number into a suitable "stem" and "leaf." In the example presented here, all the leaves are zero simply because ages were measured only to the nearest year. We see that the age difference ranges from –7 years (i.e., the wife is 7 years younger than the husband since we calculated the difference as husband's age minus wife's age) to 10 years with 12 age differences being "extreme" (i.e., >12 years). Leaving aside the extreme observations, the distribution looks normal, but inclusion of these values makes its appearance positively skewed.

If we are satisfied that the age differences have at least an approximate normal distribution, we can proceed to apply the paired samples *t*-test using the

Analyze – Compare Means – Paired-Samples T Test...

commands from the menu bar as shown in Display 2.16. The **Paired Variables** list requires the pairs of variables that are to be compared. Selecting **Options** allows for setting the significance level of the test.

Display 2.17 shows the resulting output. As for the independent samples *t*-test, SPSS displays relevant descriptive statistics, the p-value associated with the test statistic, and a confidence interval for the mean difference. The results suggest that there is a significant difference in the mean ages at marriage of men and women in married couples ($t(99)$ = 3.8, $p < 0.001$). The confidence interval indicates the age difference is somewhere between one and three years.

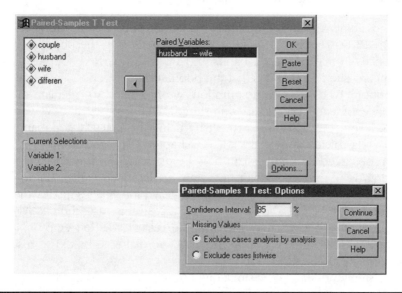

Display 2.16   Generating a paired samples *t*-test.

Paired Samples Statistics

		Mean	N	Std. Deviation	Std. Error Mean
Pair 1	husbands' ages at marriage	33.0800	100	12.31053	1.23105
	wives' ages at marriage	31.1600	100	11.00479	1.10048

Paired Samples Correlations

		N	Correlation	Sig.
Pair 1	husbands' ages at marriage & wives' ages at marriage	100	.912	.000

Paired Samples Test

		Paired Differences					t	df	Sig.(2-tailed)
					95% Confidence Interval of the Difference				
		Mean	Std. Deviation	Std. Error Mean	Lower	Upper			
Pair 1	husbands' ages at marriage - wives' ages at marriage	1.9200	5.04661	.50466	.9186	2.9214	3.805	99	.000

Display 2.17   Paired samples *t*-test output for ages at marriage.

If we are not willing to assume normality for the age differences, we can employ a nonparametric procedure such as Wilcoxon's signed ranks test (see Box 2.2) rather than the paired samples *t*-test. The test can be employed by first choosing

Analyze – Nonparametric Tests – Two Related Samples...

from the menu bar and then completing the resulting dialogue box as shown in Display 2.18.

Display 2.19 shows the resulting SPSS output. Wilcoxon's signed ranks test ranks the absolute value of the differences and then calculates the sum of the ranks within the group of pairs that originally showed positive or negative differences. Here, 67 couples showed negative differences (which according to the legend of the first table means that wives' ages were younger than husbands' ages, since SPSS calculated the differences as wife's age minus husband's age), 27 positive differences and six couples had no age difference. The negative differences have larger mean ranks than the positive differences. Consistent with the paired-samples *t*-test, the test shows that the mean age difference of men and women in married couples is not zero ($z = 3.6$, $p < 0.001$).

The normal approximation for the Wilcoxon test is appropriate when the number of pairs is large and the difference values can be ranked uniquely. In our example, the ages were recorded in full years and there are many ties in the difference variable — for example, in 20 couples, the

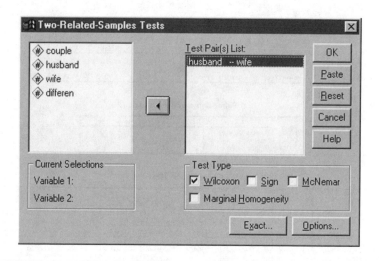

Display 2.18   Generating a Wilcoxon signed ranks test.

Ranks

		N	Mean Rank	Sum of Ranks
wives' ages at marriage - husbands' ages at marriage	Negative Ranks	67[a]	47.52	3184.00
	Positive Ranks	27[b]	47.44	1281.00
	Ties	6[c]		
	Total	100		

a. wives' ages at marriage < husbands' ages at marriage

b. wives' ages at marriage > husbands' ages at marriage

c. husbands' ages at marriage = wives' ages at marriage

Test Statistics[b]

	wives' ages at marriage - husbands' ages at marriage
Z	-3.605[a]
Asymp. Sig. (2-tailed)	.000

a. Based on positive ranks.

b. Wilcoxon Signed Ranks Test

Display 2.19   Wilcoxon's signed ranks test output for ages at marriage.

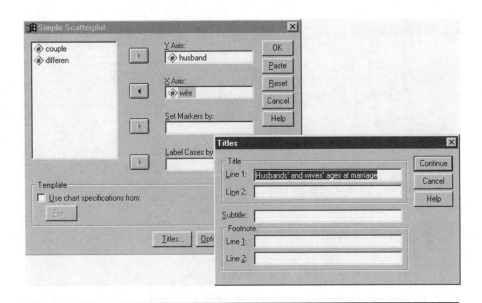

Display 2.20 Generating a scatter graph.

husband is one year older than his wife. Consequently, the approximation may not be strictly valid — for a possible alternative see Exercise 2.4.5.

Last, we will exam the relationship between the ages of husbands and wives of the couples in our data and quantify the strength of this relationship. A *scatterplot* (an *xy* plot of the two variables) often provides a useful graphical display of the relationship between two variables. The plot is obtained from the menu bar by choosing

Graphs – Scatter... – Scatterplot Simple

and defining variable lists as shown in Display 2.20.

Not surprisingly, the resulting scatterplot (shown in Display 2.21) illustrates a tendency (at least in our sample) for younger women to marry younger men and vice versa. Often it is useful to enhance scatterplots by showing a simple linear regression fit for the two variables (see later). Even more useful in many cases is to add what is known as a *locally weighted regression* fit (or *Lowess* curve) to the plot. The latter allows the data to reveal the shape of the relationship between the two ages rather than to assume that this is linear; the technique is described in detail in Cleveland (1979). Here, the Lowess curve can be added to the graph by editing the initial graph and choosing

Chart – Options

Display 2.21 Scatter graph of husbands' and wives' age at marriage.

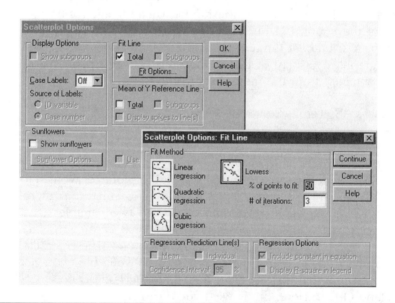

Display 2.22 Adding a Lowess curve to a scatterplot.

Display 2.23   Enhanced scatter graph resulting from commands in Display 2.22.

from the **Chart Editor** menu bar. This creates another dialogue box which needs to be filled in as shown in Display 2.22. The resulting enhanced plot is shown in Display 2.23. We shall return to this plot below.

We can quantify the relationship between ages at marriage of the husbands and wives in each couple by calculating some type of correlation coefficient. SPSS offers several under

Analyze – Correlate – Bivariate...

Here, for illustrative purposes only, we select all the coefficients available (Display 2.24). The **Variables** list contains the names of the variables that are to be correlated — where more than two variables were listed, a matrix of pairwise correlation coefficients would be generated (see Chapter 4). In this example, the resulting output, though rather lengthy, is fairly simple (Display 2.25), essentially giving the values of three correlation coefficients — the *Pearson correlation* ($r = 0.91$), *Spearman's rho* ($\rho = 0.90$), and a version of *Kendall's tau* ($\tau = 0.76$). (All these correlation coefficients are described and defined in Everitt, 2001.)

These estimates are accompanied by p-values from statistical signifi-cance tests that test the null hypothesis that the correlation in the underlying population is zero — i.e., that there is no directional relationship. Here

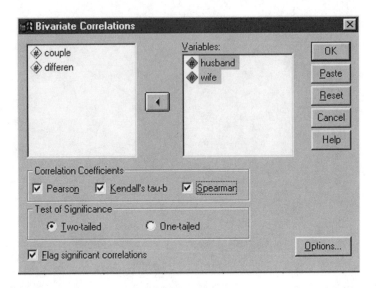

Display 2.24   Generating correlations.

we chose to consider the two-sided alternative hypothesis that there is correlation, positive or negative. For the "parametric" Pearson correlation coefficient, this test relies on the assumption of bivariate normality for the two variables. No distributional assumption is required to test Spearman's rho or Kendall's tau. But, as we would expect, the tests on each of the three coefficients indicate that the correlation between ages of husbands and wives at marriage is highly significant (the p-values of all three tests are less than 0.001).

The fitted smooth curve shown in Display 2.23 suggests that a linear relationship between husbands' age and wives' age is a reasonable assumption. We can find details of this fit by using the commands

Analyze – Regression – Linear...

This opens the **Linear Regression** dialogue box that is then completed as shown in Display 2.26. (Here husband's age is used as the dependent variable to assess whether it can be predicted from wife's age.) SPSS generates a number of tables in the output window that we will discuss in detail in Chapter 4. For now we concentrate on the table of estimated regression coefficients shown in Display 2.27.

This output table presents estimates of regression coefficients and their standard errors in the columns labeled "B" and "Std. Error," respectively.

a) "Parametric" correlation coefficients

Correlations

		husbands' ages at marriage	wives' ages at marriage
husbands' ages at marriage	Pearson Correlation	1	.912**
	Sig. (2-tailed)	.	.000
	N	100	100
wives' ages at marriage	Pearson Correlation	.912**	1
	Sig. (2-tailed)	.000	.
	N	100	100

**. Correlation is significant at the 0.01 level (2-tailed).

b) "Non-parametric" correlation coefficients

Correlations

			husbands' ages at marriage	wives' ages at marriage
Kendall's tau_b	husbands' ages at marriage	Correlation Coefficient	1.000	.761**
		Sig. (2-tailed)	.	.000
		N	100	100
	wives' ages at marriage	Correlation Coefficient	.761**	1.000
		Sig. (2-tailed)	.000	.
		N	100	100
Spearman's rho	husbands' ages at marriage	Correlation Coefficient	1.000	.899**
		Sig. (2-tailed)	.	.000
		N	100	100
	wives' ages at marriage	Correlation Coefficient	.899**	1.000
		Sig. (2-tailed)	.000	.
		N	100	100

**. Correlation is significant at the .01 level (2-tailed).

Display 2.25   Correlations between husbands' and wives' ages at marriage.

The table also gives another *t*-test for testing the null hypothesis that the regression coefficient is zero. In this example, as is the case in most applications, we are not interested in the intercept. In contrast, the slope parameter allows us to assess whether husbands' age at marriage is predictable from wives' age at marriage. The very small p-value associated with the test gives clear evidence that the regression coefficient differs from zero. The size of the estimated regression coefficient suggests that for every additional year of age of the wife at marriage, the husband's age also increases by one year (for more comments on interpreting regression coefficients see Chapter 4).

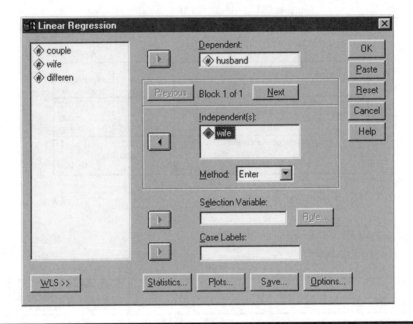

Display 2.26   Fitting a simple linear regression.

Coefficients[a]

Model		Unstandardized Coefficients		Standardized Coefficients	t	Sig.
		B	Std. Error	Beta		
1	(Constant)	1.280	1.528		.837	.404
	wives' ages at marriage	1.021	.046	.912	22.053	.000

a. Dependent Variable: husbands' ages at marriage

Display 2.27   Regression coefficient estimates for ages at marriage.

2.4   Exercises

2.4.1   Guessing the Width of a Lecture Hall

Shortly after metric units of length were officially introduced in Australia, a group of 44 students was asked to guess, to the nearest metre, the width of the lecture hall in which they were sitting. Another group of 69 students in the same room was asked to guess the width in feet, to the nearest foot (Table 2.3). (The true width of the hall was 13.1 metres or 43 feet.) Investigate by using simple graphical displays or other means which group estimated the width of the lecture hall more accurately.

Table 2.3   Guesses of the Width of a Lecture Hall

a) in meters (*n* = 44)

8	9	10	10	10	10	10	10	11	11	11	11	12	12	13
13	13	14	14	14	15	15	15	15	15	15	15	15	16	16
16	17	17	17	17	18	18	20	22	25	27	35	38	40	

b) in feet (*n* = 69)

24	25	27	30	30	30	30	30	30	32	32	33	34	34	34
35	35	36	36	36	37	37	40	40	40	40	40	40	40	40
40	41	41	42	42	42	42	43	43	44	44	44	45	45	45
45	45	45	46	46	47	48	48	50	50	50	51	54	54	54
55	55	60	60	63	70	75	80	94						

Source: Hand et al., 1994.

2.4.2 More on Lifespans of Rats: Significance Tests for Model Assumptions

In Section 2.3.1, we used several summary statistics and graphical procedures to assess the assumptions of

- Normal distributions in each group
- Homogeneity of variance

for the rats' lifespans given in Table 2.1. Use formal significance test to assess these model assumptions; specifically:

- Use *Kolmogorov-Smirnov tests* (for details, see Conover, 1998) to test the null hypotheses that the lifespans in each group follow a normal distribution.
- Use a *Levene test* (Levene, 1960) to test the null hypothesis that the variance of the lifespans is the same in both groups.

(Hint: Use the commands Analyze – Descriptive statistics – Explore and the Plots… sub-dialogue box to generate Kolmogorov-Smirnov tests and a Levene test.)

2.4.3 Motor Vehicle Theft in the U.S.

The data in Table 2.4 show rates of motor vehicle thefts in states in the U.S. (in thefts per 100,000 residents), divided into two groups depending on whether the states are located to the east or the west of the Mississippi

Table 2.4   Motor Vehicle Theft in the U.S.

Eastern State (n = 26)	Theft Rate	Western State (n = 24)	Theft Rate
Alabama	348	Alaska	565
Connecticut	731	Arizona	863
Delaware	444	Arkansas	289
Florida	826	California	1016
Georgia	674	Colorado	428
Illinois	643	Hawaii	381
Indiana	439	Idaho	165
Kentucky	199	Iowa	170
Maine	177	Kansas	335
Maryland	709	Louisiana	602
Massachusetts	924	Minnesota	366
Michigan	714	Missouri	539
Mississippi	208	Montana	243
New Hampshire	244	Nebraska	178
New Jersey	940	Nevada	593
New York	1043	New Mexico	337
North Carolina	284	North Dakota	133
Ohio	491	Oklahoma	602
Pennsylvania	506	Oregon	459
Rhode Island	954	South Dakota	110
South Carolina	386	Texas	909
Tennessee	572	Utah	238
Vermont	208	Washington	447
Virginia	327	Wyoming	149
West Virginia	154		
Wisconsin	416		

Source: Rossman, 1996. With permission of Springer-Verlag.

River. Create a box plot of theft rates for Eastern and Western states, and apply an appropriate significance test to determine whether the average rate differs in the two locations.

2.4.4 Anorexia Nervosa Therapy

Table 2.5 contains weight measurements on a group of anorexic girls who received cognitive-behavioral therapy (CBT). The weight was measured in pounds before the treatment and after 12 weeks of treatment.

Table 2.5 Weights (Pounds) of Anorexic Girls

Before Therapy	After 12 Weeks
80.5	82.2
84.9	85.6
81.5	81.4
82.6	81.9
79.9	76.4
88.7	103.6
94.9	98.4
76.3	93.4
81	73.4
80.5	82.1
85	96.7
89.2	95.3
81.3	82.4
76.5	72.5
70	90.9
80.4	71.3
83.3	85.4
83	81.6
87.7	89.1
84.2	83.9
86.4	82.7
76.5	75.7
80.2	82.6
87.8	100.4
83.3	85.2
79.7	83.6
84.5	84.6
80.8	96.2
87.4	86.7

- Assuming normality of the weight changes, use the paired samples *t*-test box to generate a 95% CI for the mean weight change over time.
- Generate the same results via a one-sample *t*-test for testing the null hypothesis that the group's mean weight change is zero.
- Assess whether there is tracking in the data, i.e., do those girls who start on a low weight also tend to have a relatively low weight after therapy?

2.4.5 More on Husbands and Wives: Exact Nonparametric Tests

Nonparametric tests for comparing groups that operate on ranks use normal approximations to generate a p-value. The approximation employed by SPSS is appropriate when the sample size is large (say $n > 25$) and there are no ties. For the Mann-Whitney U-test, the second assumption means that each observation only occurs once in the data set so that the whole sample can be ranked uniquely. For the Wilcoxon signed ranks test, this assumption implies that the absolute differences between the paired observations can be ranked uniquely. When there are only a few ties in a large sample, the normal approximation might still be used. However, when the sample is small or there are many ties, an *exact test* — a test that does not rely on approximations — is more appropriate (Conover, 1998).

SPSS provides exact Mann-Whitney U-tests and exact Wilcoxon signed ranks tests from the **Exact** sub-dialogue box of the **Two-Independent-Samples Tests** (see Display 2.10) or the **Two-Related-Samples Tests** dialogue boxes (see Display 2.18). These exact tests operate by constructing the distribution of the test statistic under the null hypothesis by permutation of the cases or pairs. The procedure is computationally intensive since it evaluates all possible permutations. SPSS sets a time limit and offers to sample a set of permutations. In the latter case, SPSS evaluates the precision of the p-value by a confidence interval.

In the ages at marriage data set (Table 2.2), the ages of husbands and wives were recorded in full years and there are many ties in the difference variable — for example, a difference value of one (the man being one year older than his wife) has been observed for 20 couples. It would, therefore, be better not to use the normal approximation previously employed in Display 2.20. Re-generate a Wilcoxon signed ranks test for the ages at marriage data — this time using an exact version of the test.

Chapter 3

Simple Inference for Categorical Data: From Belief in the Afterlife to the Death Penalty and Race

3.1 Description of Data

In this chapter, we shall consider some relatively simple aspects of the analysis of categorical data. We will begin by describing how to construct cross-classifications of raw data using the lifespan of rats and age at marriage data used in the previous chapter. Then we shall move on to examine a number of data sets in which the observations are already tabulated in this way. Details of these are as follows:

- **Belief in the afterlife by gender** — data from the 1991 General Social Survey classifying a sample of Americans according to their gender and their opinion about an afterlife (taken from Agresti, 1996, with permission of the publisher). The data appear in Table 3.1.
- **Incidence of suicidal feelings in a sample of psychotic and neurotic patients** — given in Table 3.2 (taken from Everitt, 1992).

Table 3.1   Belief in the Afterlife

		Belief in Afterlife?	
		Yes	No
Gender	Female	435	147
	Male	375	134

Table 3.2   The Incidence of Suicidal Feelings in Samples of Psychotic and Neurotic Patients

		Type of Patient	
		Psychotic	Neurotic
Suicidal feelings?	Yes	1	4
	No	19	16

Table 3.3   Oral Contraceptives

		Controls	
		Oral Contraceptive Used	Oral Contraceptive Not Used
Cases	Used	10	57
	Not used	13	95

Psychiatrists studying neurotic and psychotic patients, asked each of a sample of 20 from each category about whether they suffered from suicidal feelings.

■ **Oral contraceptive use and blood clots** — data given in Table 3.3. These data arise from a study reported by Sartwell et al. (1969). The study was conducted in a number of hospitals in several large American cities. In those hospitals, all the married women identified as suffering from idiopathic thromboembolism (blood clots) over a three-year period were individually matched with a suitable control, these being female patients discharged alive from the same hospital in the same six-month time interval as the case. In addition, they were individually matched to cases by age, marital status, race, etc. Patients and controls were then asked about their use of oral contraceptives.

■ **Alcohol and infant malformation** — data from a prospective study of maternal drinking and congenital malformations (taken from Agresti, 1996, with permission of the publisher). After the first

Table 3.4   Alcohol Consumption and Infant Malformation

		Malformation	
		Present	Absent
Average number of drinks per day	0	48	17066
	<1	38	14464
	1–2	5	788
	3–5	1	126
	≥6	1	37

Table 3.5   Race and the Death Penalty in the U.S.

	Death Penalty	Not Death Penalty
White victim		
White defendant found guilty of murder	192	1320
Black defendant found guilty of murder	110	520
Black victim		
White defendant found guilty of murder	0	90
Black defendant found guilty of murder	60	970

three months of pregnancy, the women in the sample completed a questionnaire about alcohol consumption. Following childbirth, observations were recorded on the presence or absence of congenital sex organ malformations. The data are shown in Table 3.4.

■ **Death penalty verdict by defendant's race and victim's race —** data in Table 3.5 (taken from Everitt, 1999). The data were collected to try to throw light on whether there is racial equality in the application of the death penalty in the U.S.

3.2 Methods of Analysis

Contingency tables are one of the most common ways to summarize observations on two categorical variables. Tables 3.1, 3.2, and 3.3 are all examples of 2×2 contingency tables (although Table 3.3 arises in quite a different way as we shall discuss later). Table 3.4 is an example of a 5×2 contingency table in which the row classification is ordered. Table 3.5 consists essentially of two, 2×2 contingency tables. For all such tables, interest generally lies in assessing whether or not there is any relationship or *association* between the row variable and the column variable that make up the table. Most commonly, a *chi-squared test of independence* is used to answer this question, although alternatives such as *Fisher's exact test* or *McNemar's test* may be needed when the sample size is small (Fisher's test) or the data consist of matched samples (McNemar's test). In addition, in 2 × 2 tables, it may be required to calculate a confidence interval for the ratio of population proportions. For a series of 2 × 2 tables, the *Mantel-Haenszel test* may be appropriate (see later). (Brief accounts of each of the tests mentioned are given in Box 3.1. A number of other topics to be considered later in the chapter are described in Box 3.2.)

Box 3.1 Tests for Two-Way Tables

A general 2 × 2 contingency table can be written as

		Variable 2 Category		
		1	2	Total
Variable 1	1	a	b	a + b
category	2	c	d	c + d
Total		a + c	b + d	a + b + c + d = N

(1) *Chi-squared test for a 2 × 2 contingency table*

◼ To test the null hypothesis that the two variables are independent, the relevant test statistic is

$$X^2 = \frac{N(ad - bc)^2}{(a + b)(c + d)(a + c)(b + d)}$$

◼ Under the null hypothesis of independence, the statistic has an asymptotic chi-squared distribution with a single degree of freedom.

◼ An assumption made when using the chi-square distribution as an approximation to the distribution of X^2, is that the frequencies expected under independence should not be "too small." This rather vague term has historically been interpreted as meaning not less than five, although there is considerable evidence that this rule is very conservative.

(2) *Chi-squared test for an $r \times c$ category table*

■ An $r \times c$ contingency table can be written as

		Column Variable		
		1	$\cdots$ c	Total
Row	1	n_{11}	$\cdots$ n_{1c}	$n_{1\bullet}$
variable	2	n_{21}	$\cdots$ n_{2c}	$n_{2\bullet}$
	$\vdots$	$\vdots$	$\vdots$ $\vdots$	$\vdots$
	r	n_{r1}	$\cdots$ n_{rc}	$n_{r\bullet}$
Total		$n_{\bullet1}$	$\cdots$ $n_{\bullet c}$	N

■ Under the null hypothesis that the row and column classifications are independent, estimated expected values, E_{ij}, for the ijth cell can be found as

$$E_{ij} = \frac{n_{i.}n_{.j}}{N}$$

■ The test statistic for assessing independence is

$$X^2 = \sum_{i=1}^{r} \sum_{j=1}^{c} \frac{\left(n_{ij} - E_{ij}\right)^2}{E_{ij}}$$

■ Under the null hypothesis of independence, X^2 has an asymptotic chi-squared distribution with $(r - 1)(c - 1)$ degrees of freedom.

(3) *Fisher's exact test for a 2×2 table*

■ The probability, P, of any particular arrangement of the frequencies a, b, c, and d in a 2×2 contingency table, when the marginal totals are fixed and the two variables are independent, is

$$P = \frac{(a+b)!(a+c)!(c+d)!(b+d)!}{a!\,b!\,c!\,d!\,N!}$$

■ This is known as a *hypergeometric distribution* (see Everitt, 2002a).
■ Fisher's exact test employs this distribution to find the probability of the observed arrangement of frequencies and of every arrangement giving as much or more evidence of a departure from independence, when the marginal totals are fixed.

(4) *McNemar's test*

■ The frequencies in a matched samples data set can be written as

		Sample 1		Total
		Present	Absent	
Sample 2	Present	a	b	a + b
	Absent	c	d	c + d
	Total	a + c	b + d	a + b + c + d = N

where N, in this case, is the total number of pairs.
■ Under the hypothesis that the two populations do not differ in their probability of having the characteristic present, the test-statistic

$$X^2 = \frac{(b - c)^2}{b + c}$$

has an asymptotic chi-squared distribution with a single degree of freedom.
■ An exact test can be derived by employing the Binomial distribution (see Hollander and Wolfe, 1999).

(5) *Mantel-Haenszel test*

■ For a series of k 2×2 contingency tables, the Mantel-Haenszel statistic for testing the hypothesis of no association is

$$X^2 = \frac{\left[\sum_{i=1}^{k} a_i - \sum_{i=1}^{k} \frac{(a_i + b_i)(a_i + c_i)}{N_i} \right]^2}{\sum_{i=1}^{k} \frac{(a_i + b_i)(c_i + d_i)(a_i + c_i)(b_i + d_i)}{N_i^2(N_i - 1)}}$$

where a_i, b_i, c_i, d_i represent the counts in the four cells of the *i*th table and N_i is the total number of observations in the *i*th table.

■ Under the null hypothesis, this statistic has a chi-squared distribution with a single degree of freedom.

■ The test is only appropriate if the degree and direction of the association between the two variables is the same in each stratum. A possible test of this assumption is that due to Breslow and Day (see Agresti, 1996).

(Full details of all these tests are given in Everitt, 1992.)

Box 3.2 Risk Estimates from a 2 × 2 Table

2 × 2 contingency tables often arise in medical studies investigating the relationship between possible "risk factors" and the occurrence of a particular medical condition; a number of terms arise from their use in this context that we will define here so that they can be used in the text.

(1) *Relative risk*

■ The risk of Variable 2 falling into category "1" when Variable 1 takes category "1" is estimated by $r_1 = \dfrac{a}{a+b}$

■ Similarly, the risk of Variable 2 falling into category "1" when Variable 1 takes category "2" is estimated by $r_2 = \dfrac{c}{c+d}$

■ The relative risk (RR) of Variable 2 falling into category "1" comparing Variable 1 category "1" with category "2" is estimated by RR $= \dfrac{a/(a+b)}{c/(c+d)}$

■ The relative risk of getting a disease comparing individuals exposed to a risk factor with nonexposed individuals can only be estimated from a *prospective study* (see Dunn and Everitt, 1993).

(2) *Odds ratio*

- The odds of Variable 2 falling into category "1" when Variable 1 takes category "1" is estimated by $o_1 = \dfrac{a}{b}$
- Similarly, the odds of Variable 2 falling into category "1" when Variable 1 takes category "2" is estimated by $o_2 = \dfrac{c}{d}$
- The odds ratio (OR) of Variable 2 falling into category "1" comparing Variable 1 category "1" with category "2" is estimated by $\mathrm{OR} = \dfrac{a/b}{c/d}$
- The odds ratio of a disease comparing individuals exposed to a risk factor with nonexposed individuals can be estimated from both prospective and *retrospective studies*. (Again, see Dunn and Everitt, 1993.)
- When the disease is rare (say, less than 5% of the population are affected), the population RR can be approximated by the OR (Dunn and Everitt, 1993).

3.3 Analysis Using SPSS

It is common in many areas of research for continuous variables to be converted to categorical variables by grouping them, prior to analysis, into two or more categories, although it is well known that this is not always a wise procedure (see Altman, 1998). In the next two sections we use continuous variables from the data sets introduced in the previous chapter to illustrate how continuous variables can be categorized in SPSS and how, once categorical outcomes are available, these can be easily summarized using cross-tabulations.

3.3.1 Husbands and Wives Revisited

A fairly broad categorization of the ages at marriage (Table 2.2) is given by the three categories "Early marriage" (before the age of 30 years), "Marriage in middle age" (at an age between 30 and 49 years inclusive), and "Late marriage" (at the age of 50 or later). The Recode command in SPSS is useful for creating such a grouping. To categorize both age variables, we can use the following steps:

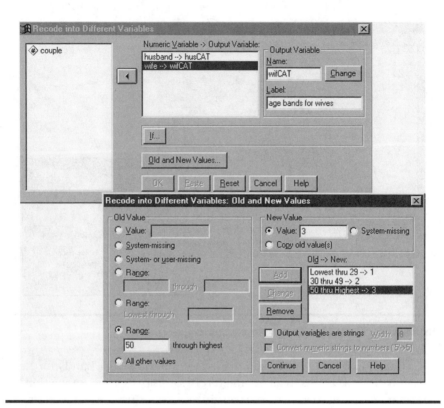

Display 3.1   Recoding variables.

- Access the commands Transform – Recode – Into Different Variables....
- Fill in the resulting dialogue box as indicated in Display 3.1 (we have labeled the new categorical variables).
- Click the button Old and New Values... to open a sub-dialogue box.
- Use this box to map ranges of the original variables onto integer numbers as indicated in Display 3.1.
- Click Continue and OK to execute the recoding.

This procedure results in two new variables, husCAT and wifCAT, being added to the Data View spreadsheet. For each couple, the spreadsheet now shows the husband's and the wife's age at marriage, as well as the corresponding categories. (To ensure that later analyses will contain explanations of the categories rather than simply print category codes, it is a good idea to always label the category codes using the Variable View.)

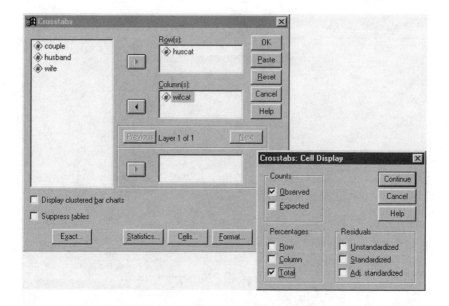

Display 3.2   Cross-classifying two categorical variables.

Cross-tabulation serves to summarize the marriage age bands for both husbands and wives as well as their relationship. A cross-tabulation of the categorized age data can be constructed by using the commands

Analyze – Descriptive Statistics – Crosstabs...

This opens the dialogue box shown in Display 3.2. We need to specify which variable is to define the rows and which the columns of the cross-tabulation. Here we choose the age bands for husbands to define the rows of the table and those of the wives the columns. Clicking the **Cells...** button opens the **Cell Display** sub-dialogue box; this controls what figures will be shown for each cell of the constructed cross-classification. Selecting the observed absolute frequencies of each cell as well as the frequencies as a percentage of the total number of couples in the study leads to the output table shown in Display 3.3.

The cross-tabulation displays the frequency distribution of the age bands for husbands and wives in the last column and row, respectively. As seen previously (e.g., Chapter 2, Display 2.14), the majority of the married people married before the age of 30. The association between husbands' and wives' ages at marriage is also apparent. Since 48% + 27% + 8% = 83% of couples, husband and wife were in the same age band at the time of marriage. The empty cells in the table show that "older" wives did not get married to "younger" men and that men older than 50 years

age bands for husbands * age bands for wives Crosstabulation

			age bands for wives			
			younger than 30 years	30 to 49 years	50 years or older	Total
age bands for husbands	younger than 30 years	Count	48	4		52
		% of Total	48.0%	4.0%		52.0%
	30 to 49 years	Count	6	27		33
		% of Total	6.0%	27.0%		33.0%
	50 years or older	Count		7	8	15
		% of Total		7.0%	8.0%	15.0%
Total		Count	54	38	8	100
		% of Total	54.0%	38.0%	8.0%	100.0%

Display 3.3   Cross-classification of age bands for wives and husbands.

diet * Lifespan less than 2 years? Crosstabulation

			Lifespan less than 2 years?		
			Yes	No	Total
diet	Restricted diet	Count	17	89	106
		% within diet	16.0%	84.0%	100.0%
	Ad libitum diet	Count	52	37	89
		% within diet	58.4%	41.6%	100.0%
Total		Count	69	126	195
		% within diet	35.4%	64.6%	100.0%

Display 3.4   Cross-tabulation of lifespan by diet.

did not get married to wives more than 20 years their juniors (at least in Cumberland County).

3.3.2 Lifespans of Rats Revisited

The lifespans of rats data set (Table 2.1) already contains one categorical outcome — diet. We will now also recode the continuous lifespan variable into a further categorical variable labeled "Lifespan less than 2 years?" with two categories (1 = "Yes" and 2 = "No"), again by using the recode command. Diet and the categorized lifespan variable can then be cross-classified as in the previous example.

Since the size of the diet groups was under the control of the investigator in this example, it made sense to express the cell counts as percentages within diet groups. So we assign diet categories to the rows of our cross-classification and check **Row** in the **Cell Display** sub-dialogue box (see Display 3.2). The resulting cross-tabulation is shown in Display 3.4. We see that 84% of the rats within the restricted diet group lived for at least two years, while a lower percentage (41.6%) reached

Risk Estimate

	Value	95% Confidence Interval	
		Lower	Upper
Odds Ratio for diet (Restricted diet / Ad libitum diet)	.136	.070	.265
For cohort Lifespan less than 2 years? = 1.00	.274	.172	.439
For cohort Lifespan less than 2 years? = 2.00	2.020	1.557	2.619
N of Valid Cases	195		

Display 3.5   Risk output for lifespan by diet cross-tabulation in Display 3.4.

this age in the *ad libitum* group. Expressed as a risk ratio, this amounts to a relative risk of dying before 2 years of age comparing the restricted diet with the *ad libitum* diet of $(17/106) : (52/89) = 0.274$. In other words, for rats on a restricted diet, the risk of dying early is about a quarter that of rats on an *ad libitum* diet.

For 2 × 2 tables, such as the one in Display 3.4, relative risk (RR) and odds ratio (OR) estimates (for definitions see Box 3.2), together with corresponding confidence intervals (CIs) can be obtained in SPSS using the steps:

■ Click the **Statistics** button on the **Crosstabs** dialogue box (see Display 3.2).
■ Check **Risk** on the resulting sub-dialogue box.

Executing these commands adds a further table to the output. This "Risk Estimate" table for the lifespan by diet cross-tabulation is shown in Display 3.5.

To understand the OR and RRs presented in this table, one needs to be aware of the comparisons used by SPSS. By default (and we are not aware of a way of changing this), SPSS presents one OR followed by two RRs. The OR compares the odds of the category corresponding to the first column of the cross-tabulation, between the categories represented by the first and second row. So, for the cross-tabulation shown in Display 3.4, the odds of living less than two years are being compared between the restricted diet group and the *ad libitum* group (estimated OR = 0.14, 95% CI from 0.07 to 0.265). The first RR value is the same comparison but this time comparing risks rather than odds (estimated RR = 0.274, 95% CI from 0.172 to 0.439). The last RR compares the risk of the category

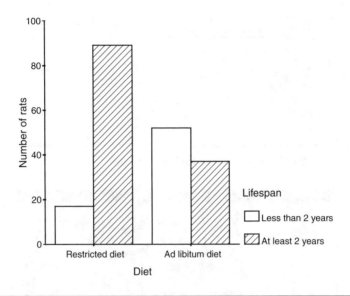

Display 3.6   Clustered bar chart for lifespan by diet cross-tabulation.

represented by the second column between the first and the second row of the table. For the cross-tabulation in Display 3.4, this means that the "risk" of living at least 2 years is being compared (estimated RR = 2.02, 95% CI from 1.557 to 2.619).

In practice these SPSS default settings mean that the groups that are to be compared should define the rows of the table. Since the first row is always compared with the second and SPSS automatically puts the category with the smallest code into the first row, the categorical variable codes will have to be chosen accordingly. In this application, for example, we are limited to comparing the restricted diet group (which was coded 1) with the *ad libitum* group (coded 2) and not vice versa. The first column (lowest category code) defines the category for which the risk or odds are of interest. Again, at least for ORs, it is advisable to code the data so that the category of interest receives the lowest code.

It is sometimes helpful to display frequency tables graphically. One possibility is to check **Display clustered bar charts** in the **Crosstabs** dialogue box (see Display 3.2). This generates a clustered bar chart with the rows of the cross-classification defining the clusters and the columns defining the patterns. Display 3.6 shows this type of plot for the cross-tabulation in Display 3.4 (after editing the appearance a little for presentation).

Now, we move on to consider the data sets introduced earlier that actually arise in the form of contingency tables.

	count	gender	belief	var
1	435	1	1	
2	147	1	2	
3	375	2	1	
4	134	2	2	
5				

Display 3.7   Data View **spreadsheet for data in Table 3.1.**

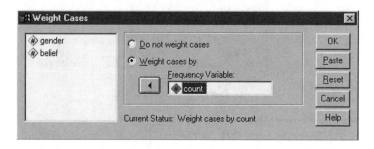

Display 3.8   **Replicating cases according to a frequency variable in SPSS.**

3.3.3 Belief in the Afterlife

We start by importing the cross-classification in Table 3.1 into SPSS. To reproduce the table in SPSS, we need to define three variables, two relating to the two categorical outcomes of interest, gender and belief, and a further variable, here called count, which holds the cell counts. The table translates into a **Data View** spreadsheet as shown in Display 3.7. Each cell of the table occupies a row in the spreadsheet and the cell count is accompanied by category codes for its corresponding row and column. (The categorical variables are each labeled with their category codes as in Table 3.1 to facilitate interpretation of later results.)

We now need to specify the true structure of the data specified in Display 3.7; otherwise, it will be interpreted by SPSS as a data set containing only four observations. For this, we need the commands

Data – Weight Cases …

This opens the dialogue box shown in Display 3.8. Checking **Weight cases by** allows the user to replicate each cell *k* times where *k* is determined by the **Frequency Variable**, here the variable count. Note that using the **Weight Cases** command does not alter the appearance of the spreadsheet.

Gender * Belief in afterlife? Crosstabulation

			Belief in afterlife?		
			Yes	No	Total
Gender	Female	Count	435	147	582
		% within Gender	74.7%	25.3%	100.0%
	Male	Count	375	134	509
		% within Gender	73.7%	26.3%	100.0%
Total		Count	810	281	1091
		% within Gender	74.2%	25.8%	100.0%

Display 3.9   Cross-tabulation of categorical variables in Display 3.7.

We can now use the **Crosstabs** command as before to generate a cross-tabulation of the categorical outcomes **belief** and **gender**. Using gender to define the rows and choosing to express counts as row percentages results in the table shown in Display 3.9.

The row percentages show that 74.7% of the females and 73.7% of the males in the study believe in an afterlife. Does this observed gender difference in our sample indicate a population association between gender and beliefs about an afterlife? We can address this question by the chi-squared test for independence, i.e., no association (for details see Box 3.1). To apply this test, we need to check **Chi-square** in the **Statistics** sub-dialogue box (Display 3.10). This adds another table to the output — see Display 3.11.

The first line of the "Chi-Square Tests" table gives the results from the chi-squared test of independence. Also supplied is a *continuity corrected* version of this test (for details, see Everitt, 1992) and a *likelihood ratio test* of independence derived from a particular statistical model (for details, see again Everitt, 1992). The remaining tests will be discussed in subsequent sections.

The standard chi-squared test of independence gives no evidence of an association between gender and beliefs about the afterlife ($X^2(1) = 0.16$, $p = 0.69$); use of the continuity corrected version or the likelihood ratio test does not change this conclusion. Consequently, to summarize beliefs about the afterlife we might simply use the marginal distribution in Display 3.9, the sex of respondents can be ignored; 74.2% of the respondents believed in an afterlife, while the remainder did not.

3.3.4 Incidence of Suicidal Feelings

The cross-tabulation of suicidal feelings by patient group shown in Table 3.2 can be converted into a **Data View** spreadsheet by following the same coding and weighting steps as in the previous example. And a cross-

Display 3.10   Generating a chi-squared test for a cross-tabulation.

Chi-Square Tests

	Value	df	Asymp. Sig. (2-sided)	Exact Sig. (2-sided)	Exact Sig. (1-sided)
Pearson Chi-Square	.162[b]	1	.687		
Continuity Correction[a]	.111	1	.739		
Likelihood Ratio	.162	1	.687		
Fisher's Exact Test				.729	.369
Linear-by-Linear Association	.162	1	.687		
N of Valid Cases	1091				

a. Computed only for a 2x2 table

b. 0 cells (.0%) have expected count less than 5. The minimum expected count is 131.10.

Display 3.11   Chi-squared test for the cross-tabulation in Display 3.9.

classification can again be created using the **Crosstabs** box. For these data, we will transpose the original table, since we want to display the table as a stacked bar chart with stacks representing the equal-sized groups, and define row categories by patient type. The SPSS steps necessary are:

Patient type * Suicidal feelings? Crosstabulation

			Suicidal feelings?		Total
			Yes	No	
Patient type	Psychotic	Count	1	19	20
		Expected Count	2.5	17.5	20.0
		% within Patient type	5.0%	95.0%	100.0%
		Residual	-1.5	1.5	
		Std. Residual	-.9	.4	
	Neurotic	Count	4	16	20
		Expected Count	2.5	17.5	20.0
		% within Patient type	20.0%	80.0%	100.0%
		Residual	1.5	-1.5	
		Std. Residual	.9	-.4	
Total		Count	5	35	40
		Expected Count	5.0	35.0	40.0
		% within Patient type	12.5%	87.5%	100.0%

Display 3.12   Cross-tabulation of suicidal feelings by patient group.

- Check Display Clustered bar chart in the Crosstabs dialogue box.
- Check Chisquare in the Statistics sub-dialogue box (see Display 3.10).
- Check Observed Counts, Expected Counts, Unstandardized Residuals, Standardized Residuals, and Row Percentages in the Cell Display sub-dialogue box (see Display 3.2).

The resulting 2 × 2 table is shown in Display 3.12. In response to the chart command, SPSS initially produces a clustered bar chart; this can easily be turned into a stacked bar chart by using the commands

Gallery – Bar... – Stacked – Replace

in the Chart Editor. The final bar chart is shown in Display 3.13.

The table and chart show that the observed percentage of patients with suicidal feelings in the sample of neurotic patients (20%) is far higher than in the sample of psychotic patients (5%). An alternative way of inspecting the departures from independence is to compare the counts observed in the cells of the table with those expected under independence of the row and column variables (for details on the calculation of expected counts, see Box 3.1). The difference is called the *residual*, with positive values indicating a higher cell frequency than expected. Here we again see that suicidal feelings are over-represented in the neurotic patients. More useful than the raw residuals are *standardized residuals* (see Everitt, 1992) that are scaled so that they reflect the contribution of each cell to the test statistic of the chi-squared test; consequently, they are useful for

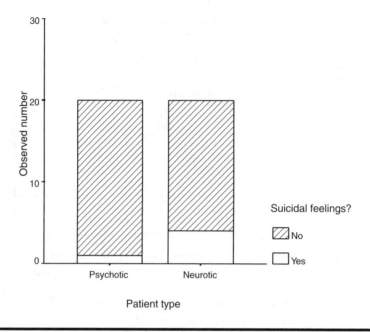

Display 3.13   Stacked bar chart of suicidal feelings within patient types.

evaluating the relative contribution of cells to an observed association between row and column variables.

As in the previous example, we are interested in testing whether this observed association between row and column variables represents evidence for an association in the underlying population. But, in contrast to the previous data set, the standard chi-squared test is perhaps not entirely appropriate since the cell counts expected under independence of suicidal feelings and patient group may be too small to justify the chi-squared approximation (see Box 3.1). Display 3.12 shows that only 5 people reported suicidal feelings in this study and, as a result, the expected counts in the cells of the first column are rather small. In fact, SPSS warns about this in the chi-squared test output (Display 3.14).

A suitable alternative test of independence in this situation is Fisher's exact test. This test does not use an approximation to derive the distribution of a test statistic under the null hypothesis; rather the exact distribution of the cell frequencies given the marginal totals is employed (for details, see Box 3.1). For 2 × 2 tables, SPSS automatically supplies Fisher's exact test when **Chisquare** is checked in the **Statistics** sub-dialogue box. (Exact tests for larger tables are also possible; see Exercise 3.4.2.)

The test output simply consists of the exact p-values for a two-sided and a one-sided test, since the test statistic is the observed cell arrangement

Chi-Square Tests

	Value	df	Asymp. Sig. (2-sided)	Exact Sig. (2-sided)	Exact Sig. (1-sided)
Pearson Chi-Square	2.057[b]	1	.151		
Continuity Correction[a]	.914	1	.339		
Likelihood Ratio	2.185	1	.139		
Fisher's Exact Test				.342	.171
Linear-by-Linear Association	2.006	1	.157		
N of Valid Cases	40				

[a]. Computed only for a 2x2 table

[b]. 2 cells (50.0%) have expected count less than 5. The minimum expected count is 2.50.

Display 3.14 **Fisher's exact test for the cross-tabulation in Display 3.12.**

Cases: Contraceptives used? * Matched controls: Contraceptives used? Crosstabulation

			Matched controls: Contraceptives used?		Total
			Yes	No	
Cases: Contraceptives used?	Yes	Count	10	57	67
		% of Total	5.7%	32.6%	38.3%
	No	Count	13	95	108
		% of Total	7.4%	54.3%	61.7%
Total		Count	23	152	175
		% of Total	13.1%	86.9%	100.0%

Display 3.15 **Cross-tabulation of contraceptive use in cases and their matched controls.**

(Display 3.14). Here we carry out a two-sided test, since we have no prior opinion regarding the direction of a possible group difference in the proportion of having suicidal feelings. At the 5% significance level, Fisher's exact test finds no evidence of an association between patient type and suicidal feelings ($p = 0.34$).

3.3.5 Oral Contraceptive Use and Blood Clots

The cross-tabulation of contraceptive use in women with idiopathic thromboembolism and their matched controls shown in Table 3.3, is first entered into SPSS using the same steps as in the previous two examples. The resulting output table, this time showing percentages of the total number of matched pairs, is shown in Display 3.15.

Chi-Square Tests

	Value	Exact Sig. (2-sided)
McNemar Test		.000[a]
N of Valid Cases	175	

a. Binomial distribution used.

Display 3.16   McNemar test for the cross-tabulation in Display 3.15.

At first sight, this table appears similar to those analyzed in the previous two sections. However, this cross-tabulation differs from the previous ones in some important aspects. Previously, we tabulated two different categorical variables that were measured on the same set of subjects. Now, we are considering a cross-tabulation of the same outcome (contraceptive use) but measured twice (once on patients and once on their matched controls). As a result, the total cell count ($N = 175$) represents the number of *matched pairs* rather than the number of individuals taking part in the study.

The different nature of this cross-tabulation has implications for the questions that might be of interest. In the previous cross-tabulations, we tested for an association between the row and the column variables. In a 2×2 table, this simply amounts to testing the experimental hypothesis that the percentage of one category of the first variable differs between the categories of the second variable. Such questions are not of interest for the current cross-tabulation. (We would be testing whether the percentage of matched controls using contraceptives is the same in the patients who use contraceptives and those who do not!) Instead, we are interested in assessing whether the probability of the potential risk factor, contraceptive use, differs between patients and their matched controls. In terms of the table in Display 3.15, it is the marginal totals (38.3% and 13.1% contraceptive use, respectively, for our samples of patients and controls) that are to be compared.

The McNemar test assesses the null hypothesis that the population marginal distributions do not differ and can be generated for a cross-tabulation by checking **McNemar** in the **Statistics** sub-dialogue box (see Display 3.10). The resulting output table is shown in Display 3.16. SPSS provides a p-value for an exact test (see Box 3.1). The two-sided test suggests that women suffering from blood clots reported a significantly higher rate of contraceptive use than controls ($p < 0.001$).

Alternatively, a McNemar test can be generated in SPSS by using the commands:

- Analyze – Nonparametric Tests – 2 Related Samples ...
- Defining the variable pair to be tested in the resulting dialogue box.
- And checking **McNemar** (see Chapter 2, Display 2.18).

This automatically provides the 2 × 2 cross-tabulation, the test statistic from a continuity corrected chi-squared test, and the resulting approximate p-value (see Box 3.1).

3.3.6 Alcohol and Infant Malformation

We now move on to the data on maternal drinking and congenital malformations in Table 3.4. Again, interest centers on the possible association between alcohol consumption and presence of infant malformation. But in this case, a more powerful test than the standard chi-squared test is available because of the ordered nature of the row variable. Here, the null hypothesis of independence can be tested against a more restricted alternative, namely, that the proportion of malformations changes gradually with increasing consumption. The *linear trend test* for a cross-tabulation assesses independence against this alternative hypothesis (for details, see Everitt, 1992).

The data in Table 3.4 can be imported into SPSS as described for the previous three examples. However, since here we want to make use of the ordering of the alcohol categories, we need to ensure that the coding of these categories reflects their order. Consequently, we assign the smallest code to the smallest consumption level, etc. Display 3.17 shows the cross-tabulation created via the **Crosstabs** dialogue box. We display row percentages because we are looking for a gradual change in these percentages over consumption levels.

The table shows that in our sample, the proportion of malformations increases with each increase in consumption level. However, the formal trend test, which is automatically generated when **Chi-square** is checked

CONSUMPT * MALFORMA Crosstabulation

			MALFORMA		
			Absent	Present	Total
CONSUMPT	0	Count	17066	48	17114
		% within CONSUMPT	99.7%	.3%	100.0%
	less than 1	Count	14464	38	14502
		% within CONSUMPT	99.7%	.3%	100.0%
	1 to 2	Count	788	5	793
		% within CONSUMPT	99.4%	.6%	100.0%
	3 to 5	Count	126	1	127
		% within CONSUMPT	99.2%	.8%	100.0%
	6 or more	Count	37	1	38
		% within CONSUMPT	97.4%	2.6%	100.0%
Total		Count	32481	93	32574
		% within CONSUMPT	99.7%	.3%	100.0%

Display 3.17 **Cross-tabulation of infant malformation by alcohol consumption.**

Chi-Square Tests

	Value	df	Asymp. Sig. (2-sided)	Exact Sig. (2-sided)	Exact Sig. (1-sided)	Point Probability
Pearson Chi-Square	12.082[a]	4	.017	.034		
Likelihood Ratio	6.202	4	.185	.133		
Fisher's Exact Test	10.458			.033		
Linear-by-Linear Association	1.828[b]	1	.176	.179	.105	.028
N of Valid Cases	32574					

a. 3 cells (30.0%) have expected count less than 5. The minimum expected count is .11.

b. The standardized statistic is 1.352.

Display 3.18   Linear trend test for cross-tabulation in Display 3.17.

in the Statistics sub-dialogue box and shown under the "Linear-by-Linear Association" heading in the "Chi-Square Tests" output table (Display 3.18), suggests that this is unlikely to reflect a linear trend in the population ($X^2(1) = 1.83$, $p = 0.18$).

3.3.7 Death Penalty Verdicts

The data shown in Table 3.5 on race and the death penalty in the U.S. are a three-way classification of the original observations. For each trial with a positive murder verdict, three categorical outcomes are recorded: the race of the defendant, the race of the victim, and use of the death penalty. The question of most interest here is: "Is there an association between the race of the defendant found guilty of murder and the use of the death penalty?" (This three-way classification and higher order examples can be imported into SPSS by using the coding and weighting steps employed in the previous examples. The only difference is that further categorical variables need to be coded.)

For illustrative purposes, we start by ignoring the race of the victim and construct a two-way classification using the Crosstabs dialogue box. The result is shown in Display 3.19. The table shows that a somewhat higher percentage of white defendants found guilty of murder (12%) were sentenced with the death penalty compared with black defendants (10.2%). Testing for an association in this table in the usual way results in a chi-squared value of 2.51 with an associated p-value of 0.11, a value that provides *no* evidence of an association between the race of the defendant and the use of the death penalty.

We now repeat the analysis, this time taking account of the three-way classification by introducing the race of the victim. A cross-classification of the use of the death penalty by defendant's race within subgroups of victim's race is generated in SPSS by defining victim race layers in the Crosstabs dialogue box (Display 3.20). The new cross-tabulation (Display 3.21), which

Race of defendant found guilty of murder * Death penalty given? Crosstabulation

			Death penalty given?		
			Yes	No	Total
Race of defendant found guilty of murder	White	Count	192	1410	1602
		% within Race of defendant found guilty of murder	12.0%	88.0%	100.0%
	Black	Count	170	1490	1660
		% within Race of defendant found guilty of murder	10.2%	89.8%	100.0%
Total		Count	362	2900	3262
		% within Race of defendant found guilty of murder	11.1%	88.9%	100.0%

Display 3.19   Cross-tabulation of death penalty use by race of defendant.

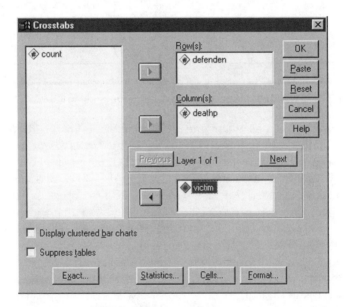

Display 3.20   Constructing a three-way classification.

replicates Table 3.5, shows higher death penalty rates for the murder of white victims than of black victims (14.1% compared with 5.4%) and within victim's race subgroups lower death penalty rates for white defendants compared with black defendants (white victims: 12.7% compared with 17.5%, black victims: 0% compared with 5.8%). These figures appear to contradict the figures obtained from Display 3.19.

The phenomenon that a measure of the association between two categorical variables (here defendant's race and death penalty) may be identical within the levels of a third variable (here victim's race), but takes

Race of defendant found guilty of murder * Death penalty given? * Race of victim Crosstabulation

Race of victim					Death penalty given?		
					Yes	No	Total
White	Race of defendant found guilty of murder	White	Count		192	1320	1512
			% within Race of defendant found guilty of murder		12.7%	87.3%	100.0%
		Black	Count		110	520	630
			% within Race of defendant found guilty of murder		17.5%	82.5%	100.0%
	Total		Count		302	1840	2142
			% within Race of defendant found guilty of murder		14.1%	85.9%	100.0%
Black	Race of defendant found guilty of murder	White	Count			90	90
			% within Race of defendant found guilty of murder			100.0%	100.0%
		Black	Count		60	970	1030
			% within Race of defendant found guilty of murder		5.8%	94.2%	100.0%
	Total		Count		60	1060	1120
			% within Race of defendant found guilty of murder		5.4%	94.6%	100.0%

Display 3.21   Cross-tabulation generated by commands in Display 3.20.

entirely different values when the third variable is disregarded and the measure calculated from the pooled table, is known as *Simpson's paradox* in the statistical literature (see Agresti, 1996). Such a situation can occur if the third variable is associated with both of the other two variables. To check that this is indeed the case here, we need to cross-tabulate victim's race with defendant's race. (We have already seen that victim's race is associated with death penalty use in Display 3.21.) The new cross-tabulation (Display 3.22) confirms that victim's race is also associated with the second variable, defendant's race, with the majority of murders being committed by a perpetrator of the same race as the victim. Therefore, the initial comparison of death penalty rates between race groups (see Display 3.19) is unable to distinguish between the effects of the defendant's race and the victim's race, and simply estimates the combined effect.

Returning now to our correct three-way classification in Display 3.21, we wish to carry out a formal test of independence between defendant's race and the use of the death penalty. Checking Chisquare in the Statistics sub-dialogue box will supply separate chi-squared tests for white and black victims, but we prefer to carry out an overall test. The Mantel-Haenszel test (for details, see Box 3.1) is the appropriate test in this situation as long as we are willing to assume that the degree and direction of a potential association between defendant's race and death penalty is the same for black and white victims.

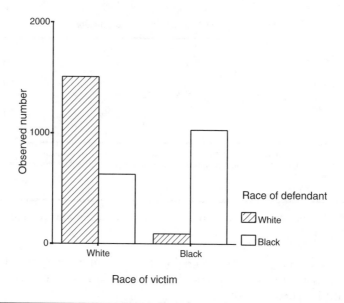

Display 3.22   Clustered bar chart for cross-tabulation of defendant's race by victim's race.

SPSS supplies the Mantel-Haenszel test for a series of 2 × 2 tables after checking Cochran's and Mantel-Haenszel statistics in the Statistics sub-dialogue box (see Display 3.10). For the death penalty data, this generates the output shown in Displays 3.23 and 3.24. The second line in Display 3.23 gives the result of the Mantel-Haenszel test; with a $X^2(1) = 11.19$ and an associated p-value of 0.001, this provides strong evidence of an association between the use of the death penalty and the race of the defendant

Tests for Homogeneity of the Odds Ratio

Statistics		Chi-Squared	df	Asymp. Sig. (2-sided)
Conditional	Cochran's	11.642	1	.001
Independence	Mantel-Haenszel	11.193	1	.001
Homogeneity	Breslow-Day	3.844	1	.050
	Tarone's	3.833	1	.050

Under the conditional independence assumption, Cochran's statistic is asymptotically distributed as a 1 df chi-squared distribution, only if the number of strata is fixed, while the Mantel-Haenszel statistic is always asymptotically distributed as a 1 df chi-squared distribution. Note that the continuity correction is removed from the Mantel-Haenszel statistic when the sum of the differences between the observed and the expected is 0.

Display 3.23   Mantel-Haenszel test for cross-tabulation in Display 3.21.

Mantel-Haenszel Common Odds Ratio Estimate

Estimate			.642
ln(Estimate)			-.443
Std. Error of ln(Estimate)			.128
Asymp. Sig. (2-sided)			.001
Asymp. 95% Confidence Interval	Common Odds Ratio	Lower Bound	.499
		Upper Bound	.825
	ln(Common Odds Ratio)	Lower Bound	-.695
		Upper Bound	-.192

The Mantel-Haenszel common odds ratio estimate is asymptotically normally distributed under the common odds ratio of 1.000 assumption. So is the natural log of the estimate.

Display 3.24 Mantel-Haenszel common odds ratio estimator for cross-tabulation in Display 3.21.

although the result of the Breslow-Day test for homogeneity of odds ratios in the two tables throws some doubt on the assumption of the homogeneity of odds ratios.

If, however, we decide not to reject the homogeneity assumption, we can find an estimate of the assumed common odds ratio. SPSS estimates the common odds ratio of the first column in the cross-tabulation (death penalty given) comparing the first row (white defendant) with the second row (black defendant). The estimates and 95% confidence intervals are shown in Display 3.24. The Mantel-Haenszel estimator of the common odds ratio is presented in the first line (OR = 0.642), showing that the chance of death penalty sentences was reduced by 35.8% for white defendants compared with black defendants. A confidence interval for the OR is constructed on the log-scale employing asymptotic normality and hence SPSS supplies the log-OR and its standard error. The final 95% confidence interval for the OR (0.499 to 0.825) shows that a reduction in death penalty odds in whites relative to blacks between 17.5% and 50.1% is consistent with the data.

3.4 Exercises

3.4.1 Depersonalization and Recovery from Depression

A psychiatrist wishes to assess the effect of the symptom "depersonalization" on the prognosis of depressed patients. For this purpose, 23 endogenous depressed patients who were diagnosed as being "depersonalized" were matched one-to-one for age, sex, duration of illness, and certain personality variables, with 23 patients who were diagnosed as being "not depersonalized." On discharge after a course of treatment, patients were assessed

Table 3.6 Recovery from Depression after Treatment

		Depersonalized Patients	
		Recovered	*Not Recovered*
Not depersonalized	Recovered	14	5
patients	Not recovered	2	2

Source: Everitt, 1992.

as "recovered" or "not recovered" with the results shown in Table 3.6. Is depersonalization associated with recovery?

3.4.2 Drug Treatment of Psychiatric Patients: Exact Tests for Two-Way Classifications

The chi-squared test for independence between two categorical variables uses a normal approximation to generate a p-value. The approximation is appropriate when the expected value under independence is large enough in each cell (say at least 5). For 2×2 tables, this assumption can be avoided by using Fisher's exact test. However, for $r \times c$ tables, SPSS does not supply an exact test by default. Exact tests can be generated for any size cross-tabulation by permutation (e.g., for details, see Manly, 1999). SPSS provides permutation tests for all test statistics used for two-way classifications after clicking the **Exact...** button in the **Crosstabs** dialogue box and checking **Exact** or **Monte Carlo** in the resulting sub-dialogue box. The procedure is computationally intensive. SPSS sets a time limit and offers to sample a set of permutations (**Monte Carlo** option). In the latter case, the precision of the p-value is evaluated by a confidence interval.

Table 3.7 gives a cross-classification of a sample of psychiatric patients by their diagnosis and whether they were prescribed drugs for their treatment (data taken from Agresti, 1996). Display the cross-tabulation graphically. Use an exact test to assess whether diagnosis and drug

Table 3.7 Drug Treatment by Psychiatric Patient Type

		Drug Treatment	
		Drugs	*No Drugs*
Patient type	Schizophrenia	105	8
	Affective disorder	12	2
	Neurosis	18	19
	Personality disorder	47	52
	Special symptoms	0	13

treatment are independent. Estimate the odds ratio of receiving drug treatment comparing schizophrenic patients, affective disorder patients and neurosis patients with personality disorder patients each.

3.4.3 Tics and Gender

The data in Table 3.8 give the incidence of tics in three age group samples of children with learning difficulties (data taken from Everitt, 1992). Is there any evidence of an association between the incidence of tics and gender after adjusting for age? Estimate the odds ratio of experiencing tics between boys and girls.

Table 3.8   Tics in Children with Learning Disabilities

		Child Experiences Tics?	
Age Range	Gender	Yes	No
5 to 9 years	Boys	13	57
	Girls	3	23
10 to 12 years	Boys	26	56
	Girls	11	29
13 to 15 years	Boys	15	56
	Girls	2	27

3.4.4 Hair Color and Eye Color

The data in Table 3.9 show the hair color and eye color of a large number of people. The overall chi-squared statistic for independence is very large and the two variables are clearly not independent. Examine the possible reasons for departure from independence by looking at the differences between the observed and expected values under independence for each cell of the table.

Table 3.9   Hair Color and Eye Color

		Hair Color				
		Fair	Red	Medium	Dark	Black
Eye color	Light	688	116	584	188	4
	Blue	326	38	241	110	3
	Medium	343	84	909	412	26
	Dark	98	48	403	681	81

Source: Everitt, 1992.

Chapter 4

Multiple Linear Regression: Temperatures in America and Cleaning Cars

---

4.1 Description of Data

In this chapter, we shall deal with two sets of data where interest lies in either examining how one variable relates to a number of others or in predicting one variable from others. The first data set is shown in Table 4.1 and includes four variables, *sex, age, extroversion,* and *car,* the latter being the average number of minutes per week a person spends looking after his or her car. According to a particular theory, people who score higher on a measure of extroversion are expected to spend more time looking after their cars since a person may project their self-image through themselves or through objects of their own. At the same time, car-cleaning behavior might be related to demographic variables such as age and sex. Therefore, one question here is how the variables sex, age, and extroversion affect the time that a person spends cleaning his or her car.

The second data set, reported by Peixoto (1990) is given in Table 4.2, and shows the normal average January minimum temperature (in °F) along with the longitude and latitude for 56 cities in the U.S. (Average minimum temperature for January is found by adding together the daily minimum

Table 4.1   Cleaning Cars

Sex (1 = male, 0 = female)	Age (years)	Extroversion Score	Car (min)
1	55	40	46
1	43	45	79
0	57	52	33
1	26	62	63
0	22	31	20
0	32	28	18
0	26	2	11
1	29	83	97
1	40	55	63
0	30	32	46
0	34	47	21
1	44	45	71
1	49	60	59
1	22	13	44
0	34	7	30
1	47	85	80
0	48	38	45
0	48	61	26
1	22	26	33
0	24	3	7
0	50	29	50
0	49	60	54
1	49	47	73
0	48	18	19
0	29	16	36
0	58	36	31
1	24	24	71
0	21	12	15
1	29	32	40
1	45	46	61
1	28	26	45
0	37	40	42
1	44	46	57
0	22	44	34
0	38	3	26
0	24	25	47
1	34	43	42
1	26	41	44
1	26	42	59
1	25	36	27

Source: Miles and Shevlin, 2001. With permission of the publisher.

Table 4.2   Minimum Temperatures in America

City	Temperature	Latitude	Longitude
Mobile, AL	44	31.2	88.5
Montgomery, AL	38	32.9	86.8
Phoenix, AZ	35	33.6	112.5
Little Rock, AR	31	35.4	92.8
Los Angeles, CA	47	34.3	118.7
San Francisco, CA	42	38.4	123.0
Denver, CO	15	40.7	105.3
New Haven, CT	22	41.7	73.4
Wilmington, DE	26	40.5	76.3
Washington, DC	30	39.7	77.5
Jacksonville, FL	45	31.0	82.3
Key West, FL	65	25.0	82.0
Miami, FL	58	26.3	80.7
Atlanta, GA	37	33.9	85.0
Boise, ID	22	43.7	117.1
Chicago, IL	19	42.3	88.0
Indianapolis, IN	21	39.8	86.9
Des Moines, IA	11	41.8	93.6
Wichita, KS	22	38.1	97.6
Louisville, KY	27	39.0	86.5
New Orleans, LA	45	30.8	90.2
Portland, ME	12	44.2	70.5
Baltimore, MD	25	39.7	77.3
Boston, MA	23	42.7	71.4
Detroit, MI	21	43.1	83.9
Minneapolis, MN	2	45.9	93.9
St Louis, MO	24	39.3	90.5
Helena, MT	8	47.1	112.4
Omaha, NE	13	41.9	96.1
Concord, NH	11	43.5	71.9
Atlantic City, NJ	27	39.8	75.3
Albuquerque, NM	24	35.1	106.7
Albany, NY	14	42.6	73.7
New York, NY	27	40.8	74.6
Charlotte, NC	34	35.9	81.5
Raleigh, NC	31	36.4	78.9
Bismarck, ND	0	47.1	101.0
Cincinnati, OH	26	39.2	85.0
Cleveland, OH	21	42.3	82.5
Oklahoma City, OK	28	35.9	97.5
Portland, OR	33	45.6	123.2

Table 4.2 (continued)   Minimum Temperatures in America

City	Temperature	Latitude	Longitude
Harrisburg, PA	24	40.9	77.8
Philadelphia, PA	24	40.9	75.5
Charleston, SC	38	33.3	80.8
Nashville, TN	31	36.7	87.6
Amarillo, TX	24	35.6	101.9
Galveston, TX	49	29.4	95.5
Houston, TX	44	30.1	95.9
Salt Lake City, UT	18	41.1	112.3
Burlington, VT	7	45.0	73.9
Norfolk, VA	32	37.0	76.6
Seattle, WA	33	48.1	122.5
Spokane, WA	19	48.1	117.9
Madison, WI	9	43.4	90.2
Milwaukee, WI	13	43.3	88.1
Cheyenne, WY	14	41.2	104.9

Source: Hand et al., 1994.

temperatures and dividing by 31; then the January average minima for the years 1931 to 1960 are averaged over the 30 years.) Interest lies in developing an equation relating temperature to position as given by the latitude and longitude.

We shall use *multiple linear regression* to analyze each data set.

4.2 Multiple Linear Regression

Multiple linear regression is a method of analysis for assessing the strength of the relationship between each of a set of *explanatory variables* (sometimes known as *independent variables*, although this is not recommended since the variables are often correlated), and a single *response* (or *dependent*) variable. When only a single explanatory variable is involved, we have what is generally referred to as *simple linear regression*, an example of which was given in Chapter 2.

Applying multiple regression analysis to a set of data results in what are known as *regression coefficients*, one for each explanatory variable. These coefficients give the estimated change in the response variable associated with a unit change in the corresponding explanatory variable, *conditional* on the other explanatory variables remaining constant. The fit of a multiple regression model can be judged in various ways, for example, calculation of the *multiple correlation coefficient* or by the examination of *residuals*, each of which will be illustrated later. (Further details of multiple regression are given in Box 4.1.)

Box 4.1 Multiple Linear Regression

- The multiple regression model for a response variable, y, with observed values, y_1, y_2, ..., y_n (where n is the sample size) and q explanatory variables, x_1, x_2, ..., x_q with observed values, x_{1i}, x_{2i}, ..., x_{qi} for $i = 1$, ..., n, is:

$$y_i = \beta_0 + \beta_1 x_{1i} + \beta_2 x_{2i} + ... + \beta_q x_{qi} + \varepsilon_i$$

- The term ε_i is the *residual* or *error* for individual i and represents the deviation of the observed value of the response for this individual from that expected by the model. These error terms are assumed to have a normal distribution with variance σ^2.
- The latter assumption implies that, conditional on the explanatory variables, the response variable is normally distributed with a mean that is a linear function of the explanatory variables, and a variance that does not depend on these variables.
- The explanatory variables are, strictly, assumed to be fixed; that is, they are not considered random variables. In practice, where this is unlikely, the results from a regression are interpreted as being *conditional* on the observed values of the explanatory variables.
- Explanatory variables can include continuous, binary, and categorical variables, although care is needed in how the latter are coded before analysis as we shall describe in the body of the chapter.
- The regression coefficients, β_0, β_1, ···, β_q, are generally estimated by *least squares*; details are given in Draper and Smith (1998).
- The variation in the response variable can be partitioned into a part due to regression on the explanatory variables and a residual term. The latter divided by its degrees of freedom (the *residual mean square*) gives an estimate of σ^2, and the ratio of the regression mean square to the residual mean square provides an *F*-test of the hypothesis that each of β_0, β_1, ···, β_q takes the value zero.

■ A measure of the fit of the model is provided by the *multiple correlation coefficient*, R, defined as the correlation between the observed values of the response variable and the values predicted by the model; that is,

$$\hat{y}_i = \hat{\beta}_0 + \hat{\beta}_1 x_{i1} + \cdots \hat{\beta}_q x_{iq}$$

■ The value of R^2 gives the proportion of the variability of the response variable accounted for by the explanatory variables.
■ Individual regression coefficients can be assessed using the ratio $\hat{\beta}_j / SE(\hat{\beta}_j)$, judged as a Student's t-statistic, although the results of such tests should only be used as rough guides to the 'significance' or otherwise of the coefficients for reasons discussed in the text.
■ For complete details of multiple regression, see, for example, Rawlings, Pantula and Dickey (1998).

4.3 Analysis Using SPSS

4.3.1 Cleaning Cars

In the car cleaning data set in Table 4.1, each of the variables — extroversion (extrover in the Data View spreadsheet), sex (sex), and age (age) — might be correlated with the response variable, amount of time spent car cleaning (car). In addition, the explanatory variables might be correlated among themselves. All these correlations can be found from the correlation matrix of the variables, obtained by using the commands

Analyze – Correlate – Bivariate...

and including all four variables under the Variables list in the resulting dialogue box (see Chapter 2, Display 2.24). This generates the output shown in Display 4.1.

The output table provides Pearson correlations between each pair of variables and associated significance tests. We find that car cleaning is positively correlated with extroversion ($r = 0.67$, $p < 0.001$) and being male ($r = 0.661$, $p < 0.001$). The positive relationship with age ($r = 0.234$) does not reach statistical significance ($p = 0.15$). The correlations between the explanatory variables imply that both older people and men are more extroverted ($r = 0.397$, $r = 0.403$).

Correlations

		gender	age in years	extroversion	minutes per week
gender	Pearson Correlation	1	-.053	.403**	.661**
	Sig. (2-tailed)	.	.744	.010	.000
	N	40	40	40	40
age in years	Pearson Correlation	-.053	1	.397*	.234
	Sig. (2-tailed)	.744	.	.011	.147
	N	40	40	40	40
extroversion	Pearson Correlation	.403**	.397*	1	.670**
	Sig. (2-tailed)	.010	.011	.	.000
	N	40	40	40	40
minutes per week	Pearson Correlation	.661**	.234	.670**	1
	Sig. (2-tailed)	.000	.147	.000	.
	N	40	40	40	40

**. Correlation is significant at the 0.01 level (2-tailed).

*. Correlation is significant at the 0.05 level (2-tailed).

Display 4.1   Correlation matrix for variables in the car cleaning data set.

Since all the variables are correlated to some extent, it is difficult to give a clear answer to whether, for example, extroversion is really related to car cleaning time, or whether the observed correlation between the two variables arises from the relationship of extroversion to both age and sex, combined with the relationships of each of the latter two variables to car cleaning time. (A technical term for such an effect would be *confounding*.) Similarly the observed relationship between car cleaning time and gender could be partly attributable to extroversion.

In trying to disentangle the relationships involved in a set of variables, it is often helpful to calculate *partial correlation coefficients*. Such coefficients measure the strength of the linear relationship between two continuous variables that cannot be attributed to one or more confounding variables (for more details, see Rawlings, Pantula, and Dickey, 1998). For example, the partial correlation between car cleaning time and extroversion rating "partialling out" or "controlling for" the effects of age and gender measures the strength of relationship between car cleaning times and extroversion that cannot be attributed to relationships with the other explanatory variables. We can generate this correlation coefficient in SPSS by choosing

Analyze – Correlate – Partial...

from the menu and filling in the resulting dialogue box as shown in Display 4.2. The resulting output shows the partial correlation coefficient together with a significance test (Display 4.3). The estimated partial correlation between car cleaning and extroversion, 0.51, is smaller than the previous unadjusted correlation coefficient, 0.67, due to part of the

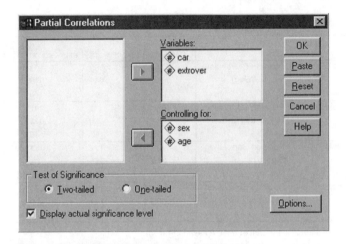

Display 4.2 Generating the partial correlation between car cleaning time and extroversion rating controlling for age and gender.

```
    - - - P A R T I A L   C O R R E L A T I O N   C O E F F I C I E N T S - - -

    Controlling for..   SEX      AGE

                        CAR     EXTROVER

    CAR             1.0000       .5107
                    (    0)     (   36)
                    P= .        P= .001

    EXTROVER          .5107    1.0000
                    (   36)     (    0)
                    P= .001    P= .

    (Coefficient / (D.F.) / 2-tailed Significance)

    " . " is printed if a coefficient cannot be computed
```

Display 4.3 Partial correlation output.

relationship being attributed to gender and/or age. We leave it as an exercise to the reader to generate the reduced partial correlation, 0.584, between car cleaning time and gender after controlling for extroversion and age.

Thus far, we have quantified the *strength* of relationships between our response variable, car cleaning time, and each explanatory variable after adjusting for the effects of the other explanatory variables. We now proceed to use the multiple linear regression approach with dependent variable, **car**, and explanatory variables, **extrover**, **sex**, and **age**, to quantify the *nature* of relationships between the response and explanatory variables after adjusting for the effects of other variables. (This is a convenient

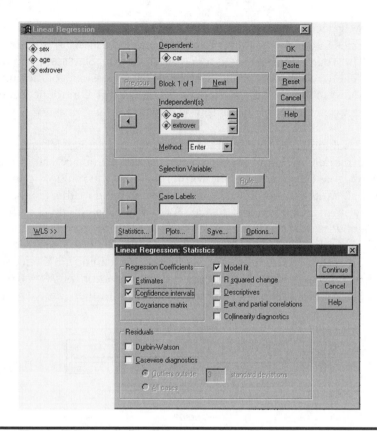

Display 4.4 Specifying a multiple linear regression model.

point to note that categorical explanatory variables, such as gender, can be used in multiple linear regression modeling as long they are represented by *dummy variables*. To "dummy-code" a categorical variable with *k* categories, *k*-1 binary dummy variables are created. Each of the dummy variables relates to a single category of the original variable and takes the value "1" when the subject falls into the category and "0" otherwise. The category that is ignored in the dummy-coding represents the *reference category*. Here **sex** is the dummy variable for category "male," hence category "female" represents the reference category.)

A multiple regression model can be set up in SPSS by using the commands

Analyze – Regression – Linear...

This results in the **Linear Regression** dialogue box shown in Display 4.4:

■ We specify the dependent variable and the set of explanatory variables under the headings **Dependent** and **Independent(s)**, respectively.
■ The regression output is controlled via the **Statistics...** button. By default, SPSS only prints estimates of regression coefficients and some model fit tables. Here we also ask for confidence intervals to be included in the output (Display 4.4).

The resulting SPSS output tables are shown in Display 4.5 and Display 4.6. The model fit output consists of a "Model Summary" table and an "ANOVA" table (Display 4.5). The former includes the multiple correlation coefficient, R, its square, R^2, and an adjusted version of this coefficient as summary measures of model fit (see Box 4.1). The multiple correlation coefficient $R = 0.799$ indicates that there is a strong correlation between the observed car cleaning times and those predicted by the regression model. In terms of variability in observed car cleaning times accounted for by our fitted model, this amounts to a proportion of $R^2 = 0.634$, or 63.4%. Since by definition R^2 will increase when further terms are added

Model Summary

Model	R	R Square	Adjusted R Square	Std. Error of the Estimate
1	.799[a]	.638	.608	13.021

a. Predictors: (Constant), age in years, gender, extroversion

ANOVA[b]

Model		Sum of Squares	df	Mean Square	F	Sig.
1	Regression	10746.290	3	3582.097	21.126	.000[a]
	Residual	6104.085	36	169.558		
	Total	16850.375	39			

a. Predictors: (Constant), age in years, gender, extroversion
b. Dependent Variable: minutes per week

Display 4.5   Multiple regression model fit output.

Coefficients[a]

Model		Unstandardized Coefficients		Standardized Coefficients			95% Confidence Interval for B	
		B	Std. Error	Beta	t	Sig.	Lower Bound	Upper Bound
1	(Constant)	11.306	7.315		1.546	.131	-3.530	26.142
	extroversion	.464	.130	.439	3.564	.001	.200	.728
	age in years	.156	.206	.085	.754	.455	-.263	.574
	gender	20.071	4.651	.489	4.315	.000	10.638	29.505

a. Dependent Variable: minutes per week

Display 4.6   Multiple regression coefficients output.

to the model even if these do not explain variability in the population, the *adjusted* R^2 is an attempt at improved estimation of R^2 in the population. The index is adjusted down to compensate for chance increases in R^2, with bigger adjustments for larger sets of explanatory variables (see Der and Everitt, 2001). Use of this adjusted measure leads to a revised estimate that 60.8% of the variability in car cleaning times in the population can be explained by the three explanatory variables.

The error terms in multiple regression measure the difference between an individual's car cleaning time and the mean car cleaning time of subjects of the same age, sex, and extroversion rating in the underlying population. According to the regression model, the mean deviation is zero (positive and negative deviations cancel each other out). But the more variable the error, the larger the absolute differences between observed cleaning times and those expected. The "Model Summary" table provides an estimate of the standard deviation of the error term (under "Std. Error of the Estimate"). Here we estimate the mean absolute deviation as 13.02 min, which is small considering that the observed car cleaning times range from 7 to 97 min per week.

Finally, the "ANOVA" table provides an *F*-test for the null hypothesis that none of the explanatory variables are related to car cleaning time, or in other words, that R^2 is zero (see Box 4.1). Here we can clearly reject this null hypothesis ($F(3,36) = 21.1$, $p < 0.001$), and so conclude that at least one of age, sex, and extroversion is related to car cleaning time.

The output shown in Display 4.6 provides estimates of the regression coefficients, standard errors of the estimates, *t*-tests that a coefficient takes the value zero, and confidence intervals (see Box 4.1). The estimated regression coefficients are given under the heading "Unstandardized Coefficients B"; these give, for each of the explanatory variables, the predicted change in the dependent variable when the explanatory variable is increased by one unit conditional on all the other variables in the model remaining constant. For example, here we estimate that the weekly car cleaning time is increased by 0.464 min for every additional score on the extroversion scale (or by 4.64 min per week for an increase of 10 units on the extroversion scale) provided that the individuals are of the same age and sex. Similarly, the estimated effect for a ten-year increase in age is 1.56 min per week. The interpretation of regression coefficients associated with dummy variables is also straightforward; they give the predicted difference in the dependent variable between the category that has been dummy-coded and the reference category. For example, here we estimate that males spend 20.07 min more per week car washing than females after adjusting for age and extroversion rating.

The regression coefficient estimate of extroversion has a standard error (heading "Std. Error") of 0.13 min per week and a 95% confidence interval for the coefficient is given by [0.200, 0.728], or in other words, the increase in car cleaning time per increase of ten in extroversion rating is estimated to be in the range 2.00 to 7.28 min per week. (Those interested in p-values can use the associated t-test to test the null hypothesis that extroversion has no effect on car cleaning times.)

Finally, the Coefficients table provides *standardized regression coefficients* under the heading "Standardized Coefficients Beta". These coefficients are standardized so that they measure the change in the dependent variable in units of its standard deviation when the explanatory variable increases by one standard deviation. The standardization enables the comparison of effects across explanatory variables (more details can be found in Everitt, 2001b). For example, here increasing extroversion by one standard deviation (SD = 19.7) is estimated to increase car cleaning time by 0.439 standard deviations (SD = 20.8 min per week). The set of beta-coefficients suggests that, after adjusting for the effects of other explanatory variables, gender has the strongest effect on car cleaning behavior.

(Note that checking **Descriptives** and **Part and partial correlations** in the **Statistics** sub-dialogue box in Display 4.4 provides summary statistics of the variables involved in the multiple regression model, including the Pearson correlation and partial correlation coefficients shown in Displays 4.1 and 4.3.)

For the car cleaning data, where there are only three explanatory variables, using the ratio of an estimated regression coefficient to its standard error in order to identify those variables that are predictive of the response and those that are not, is a reasonable approach to developing a possible simpler model for the data (that is, a model that contains fewer explanatory variables). But, in general, where a larger number of explanatory variables are involved, this approach will not be satisfactory. The reason is that the regression coefficients and their associated standard errors are estimated *conditional* on the other explanatory variables in the current model. Consequently, if a variable is removed from the model, the regression coefficients of the remaining variables (and their standard errors) will change when estimated from the data excluding this variable. As a result of this complication, other procedures have been developed for selecting a subset of explanatory variables, most associated with the response. The most commonly used of these methods are:

■ *Forward selection.* This method starts with a model containing none of the explanatory variables. In the first step, the procedure considers variables one by one for inclusion and selects the variable that results in the largest increase in R^2. In the second step, the procedures considers variables for inclusion in a model that only contains the variable selected in the first step. In each step, the variable with the largest increase in R^2 is selected until, according to an F-test, further additions are judged to not improve the model.

■ *Backward selection*. This method starts with a model containing all the variables and eliminates variables one by one, at each step choosing the variable for exclusion as that leading to the smallest decrease in R^2. Again, the procedure is repeated until, according to an F-test, further exclusions would represent a deterioration of the model.

■ *Stepwise selection*. This method is, essentially, a combination of the previous two approaches. Starting with no variables in the model, variables are added as with the forward selection method. In addition, after each inclusion step, a backward elimination process is carried out to remove variables that are no longer judged to improve the model.

Automatic variable selection procedures are exploratory tools and the results from a multiple regression model selected by a stepwise procedure should be interpreted with caution. Different automatic variable selection procedures can lead to different variable subsets since the importance of variables is evaluated relative to the variables included in the model in the previous step of the procedure. A further criticism relates to the fact that a number of tests are employed during the course of the automatic procedure, increasing the chance of false positive findings in the final model. Certainly none of the automatic procedures for selecting subsets of variables are foolproof; they must be used with care and warnings such as the following given in Agresti (1996) should be noted:

> Computerized variable selection procedures should be used with caution. When one considers a large number of terms for potential inclusion in a model, one or two of them that are not really important may look impressive simply due to chance. For instance, when all the true effects are weak, the largest sample effect may substantially overestimate its true effect. In addition, it often makes sense to include certain variables of special interest in a model and report their estimated effects even if they are not statistically significant at some level.

In addition, the comments given in McKay and Campbell (1982a, b) concerning the validity of the F-tests used to judge whether variables should be included in or eliminated from a model need to be considered.

Here, primarily for illustrative purposes, we carry out an automatic forward selection procedure to identify the most important predictors of car cleaning times out of age, sex, and extroversion, although previous results give, in this case, a very good idea of what we will find. An

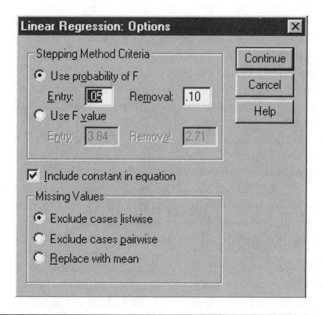

Display 4.7 Setting inclusion and exclusion criteria for automatic variable selection.

automatic forward variable selection procedure is requested from SPSS by setting the **Method** option in the **Linear Regression** dialogue box to **Forward** (see Display 4.4). When evaluating consecutive models, it is helpful to measure the change in R^2 and to consider *collinearity diagnostics* (see later), both of which can be requested in the **Statistics** sub-dialogue box (see Display 4.4). The **Options** sub-dialogue box defines the criterion used for variable inclusion and exclusion. The default settings are shown in Display 4.7. The inclusion (and exclusion) criteria can either be specified in terms of the significance level of an *F*-test (check **Use probability of F**) or in terms of a threshold value of the F-statistic (check **Use F-value**). By default, SPSS chooses a less stringent criteria for removal than for entry; although here for the automatic forward selection only, the entry criterion (significant increase in R^2 according to an *F*-test at the 5% test level) is relevant.

Display 4.8 shows the results from the automatic forward variable selection. SPSS repeats the entry criterion used and lists the variables selected in each step. With three potential predictor variables, the procedure iterated through three steps. In the first step, the variable extroversion was included. In the second step, gender was added to the model. No variable was added in the third step since the remaining potential predictor variable, age, did not improve the model according to our chosen inclusion criterion. The *F*-tests employed in each step are shown in the "Model Summary" table. Here the model selected after the first step (extroversion only) explained 44.9% of the variance in car cleaning times and the test for the single regression coefficient is highly significant ($F(1,38) = 31$, $p < 0.001$). Adding gender to the model increases the percentage variance explained by 18.3% ($F(1,37) = 18.4$, $p < 0.001$). (We leave as an exercise to check that backward and stepwise variable selection leads to the same subset of variables for this data example. But remember, this may not always be the case.)

For stepwise procedures, the "Coefficients" table shows the regression coefficients estimated for the model at each step. Here we note that the unadjusted effect of extroversion on car cleaning time was estimated to be an increase in car cleaning time of 7.08 min per week per 10 point increase on the extroversion scale. When adjusting for gender (model 2), this effect reduces to 5.09 min per week per 10 points (95% CI from 2.76 to 7.43 min per week per 10 points). SPSS also provides information about the variables not included in the regression model at each step. The "Excluded Variables" table provides standardized regression coefficients (under "Beta in") and t-tests for significance. For example, under Model 1, we see that gender, which had not been included in the model at this stage, might be an important variable since its standardized effect after adjusting for extroversion is of moderate size (0.467), there also remains moderate size partial correlation between gender and car cleaning after controlling for extroversion (0.576).

Approximate linear relationships between the explanatory variables, *multicollinearity*, can cause a number of problems in multiple regression, including:

- It severely limits the size of the multiple correlation coefficient because the explanatory variables are primarily attempting to explain much of the same variability in the response variable (see Dizney and Gromen, 1967, for an example).
- It makes determining the importance of a given explanatory variable difficult because the effects of explanatory variables are confounded due to their intercorrelations.
- It increases the variances of the regression coefficients, making use of the estimated model for prediction less stable. The parameter estimates become unreliable (for more details, see Belsley, Kuh, and Welsh, 1980).

Spotting multicollinearity among a set of explanatory variables might not be easy. The obvious course of action is to simply examine the correlations between these variables, but while this is a good initial step that is often helpful, more subtle forms of multicollinearity involving more than two variables might exist. A useful approach is the examination of the *variance inflation factors* (VIFs) or the *tolerances* of the explanatory variables. The tolerance of an explanatory variable is defined as the

Variables Entered/Removed [a]

Model	Variables Entered	Variables Removed	Method
1	extroversion	.	Forward (Criterion: Probabilit y-of-F-to-e nter <= .050)
2	gender	.	Forward (Criterion: Probabilit y-of-F-to-e nter <= .050)

a. Dependent Variable: minutes per week

Model Summary

Model	R Square Change	Change Statistics				
		F Change	df1	df2	Sig. F Change	
1	.449[a]	30.982	1	38	.000	
2	.183[b]	18.389	1	37	.000	

a. Predictors: (Constant), extroversion

b. Predictors: (Constant), extroversion, gender

Display 4.8 Automatic forward variable selection output.

Coefficients[a]

Model		Unstandardized Coefficients		Standardized Coefficients	t	Sig.	95% Confidence Interval for B		Collinearity Statistics	
		B	Std. Error	Beta			Lower Bound	Upper Bound	Tolerance	VIF
1	(Constant)	17.905	5.319		3.366	.002	7.137	28.674		
	extroversion	.708	.127	.670	5.566	.000	.451	.966	1.000	1.000
2	(Constant)	15.680	4.436		3.534	.001	6.691	24.669		
	extroversion	.509	.115	.482	4.423	.000	.276	.743	.838	1.194
	gender	19.180	4.473	.467	4.288	.000	10.118	28.243	.838	1.194

a. Dependent Variable: minutes per week

Excluded Variables[c]

Model		Beta In	t	Sig.	Partial Correlation	Collinearity Statistics		Minimum Tolerance
						Tolerance	VIF	
1	age in years	-.039[a]	-.290	.773	-.048	.842	1.187	.842
	gender	.467[a]	4.288	.000	.576	.838	1.194	.838
2	age in years	.085[b]	.754	.455	.125	.788	1.269	.662

a. Predictors in the Model: (Constant), extroversion
b. Predictors in the Model: (Constant), extroversion, gender
c. Dependent Variable: minutes per week

Display 4.8 (continued).

proportion of variance of the variable in question *not* explained by a regression on the remaining explanatory variables with smaller values indicating stronger relationships. The VIF of an explanatory variable measures the inflation of the variance of the variable's regression coefficient relative to a regression where all the explanatory variables are independent. The VIFs are inversely related to the tolerances with larger values indicating involvement in more severe relationships (according to a rule of thumb, VIFs above 10 or tolerances below 0.1 are seen as a cause of concern).

Since we asked for **Collinearity diagnostics** in the **Statistics** sub-dialogue box, the "Coefficients" table and the "Excluded Variables" table in Display 4.8 include columns labeled "Collinearity Statistics." In the "Coefficients" table, the multicollinearities involving the explanatory variables of the respective model are assessed. For example, the model selected in the second step of the procedure included extroversion and gender as explanatory variables. So a multicollinearity involving these two variables (or more simply, their correlation) has been assessed. In the "Excluded Variables" table, multicollinearities involving the excluded variable and those included in the model are assessed. For example, under "Model 2," multicollinearities involving age (which was excluded) and extroversion and gender (which were included) are measured. Here none of the VIFs give reason for concern. (SPSS provides several other collinearity diagnostics, but we shall not discuss these because they are less useful in practice than the VIFs.)

It might be helpful to visualize our regression of car cleaning times on gender and extroversion rating by constructing a suitable graphical display of the fitted model. Here, with only one continuous explanatory variable and one categorical explanatory variable, this is relatively simple since a scatterplot of the predicted values against extroversion rating can be used. First, the predicted (or *fitted*) values for the subjects in our sample need to be saved via the **Save...** button on the **Linear Regression** dialogue box (see Display 4.4). This opens the **Save** sub-dialogue box shown in Display 4.9 where **Unstandardized Predicted Values** can be requested. Executing the command includes a new variable **pre_1** on the right-hand side of the **Data View** spreadsheet.

This variable can then be plotted against the extroversion variable using the following instructions:

- The predicted value variable, **pre_1**, is declared as the **Y Axis** and the extroversion variable, **extrover** as the **X Axis** in the **Simple Scatterplot** dialogue box (see Chapter 2, Display 2.20).
- The gender variable, **sex**, is included under the **Set Markers** by list to enable later identification of gender groups.
- The resulting graph is then opened in the **Chart Editor** and the commands **Format – Interpolation... – Straight** used to connect the points.

Display 4.9   Generating predicted values.

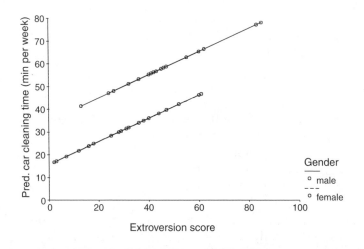

Display 4.10   Car cleaning times predicted by the multiple linear regression on gender and extroversion rating.

The final graph shown in Display 4.10 immediately conveys that the amount of time spent car cleaning is predicted to increase with extroversion rating, with the strength of the effect determined by the slope of two parallel lines (5.1 min per week per 10 points on the extroversion scale). It also shows that males are estimated to spend more time cleaning their cars with the increase in time given by the vertical distance between the two parallel lines (19.18 min per week).

4.3.2 *Temperatures in America*

For the temperature data in Table 4.2, we would like to develop an equation that allows January average temperature in a city to be predicted from spatial position as specified by its longitude and latitude. So here the dependent variable is the temperature (**temp** in our **Data View** spreadsheet) and we have two explanatory variables, latitude (**lati**) and longitude (**longi**).

To get an impression of the spatial locations of the 56 cities, we can create a scatterplot of our two explanatory variables. We plot latitude on the *y*-axis since larger values indicate more northern locations. Longitude essentially measures the distance from the Greenwich meridian, with larger values for American cities indicating more westerly locations. We therefore create a new variable using the **Compute** command (Longitude multiplied by the factor –1, called **newlongi**, see Chapter 1, Display 1.13) so that larger values indicate more easterly positions and plot this new variable on the *x*-axis.

To include city names on the scatterplot (included in the **Data View** spreadsheet as a string variable **city**), we set **Label Cases by** in the **Simple Scatterplot** dialogue box to **city**. This initially generates a scatterplot without showing the labels. But these can then be made visible as follows:

■ Open the **Chart Editor** and use the commands **Chart – Options...** from the menu bar.
■ Set **Case Labels** to **On** in the resulting Scatterplot Options box while keeping **Source of Labels** as **ID variable**.
■ To improve the appearance, we can select the city labels by highlighting them on the graph.
■ Use the **Format – Text...** commands to change the font size to 5.

The resulting scatterplot is given in Display 4.11. This demonstrates that the 56 cities in this study provide reasonable representation of the whole of the U.S.

Another graphical display that is often useful at this initial stage of analysis is the *scatterplot matrix* (see Cleveland, 1993). This is a rectangular grid arrangement of the pairwise scatterplots of the variables. In SPSS, such a plot can be produced as follows:

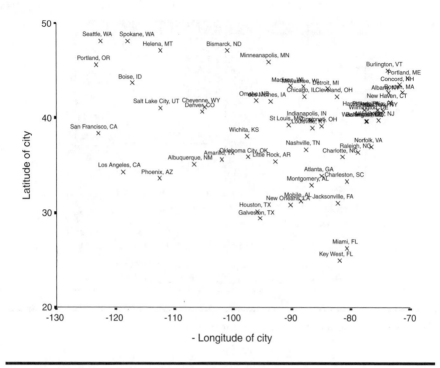

Display 4.11   Spatial locations of American cities.

- ▪ Use commands **Graph – Scatter...** from the menu bar.
- ▪ Choose **Matrix** from the resulting **Scatterplot** dialogue box.
- ▪ Select the dependent variable as well as all the explanatory variables as **Matrix Variables** in the next **Scatterplot Matrix** dialogue box as shown in Display 4.12.

Inclusion of regression lines from simple linear regressions of the data in each panel of the scatterplot matrix requires the following steps:

- ▪ Select the commands **Chart – Options...** from the **Chart editor** menu bar.
- ▪ Check **Total** on the resulting **Scatterplot Options** dialogue box.
- ▪ Click the **Fit Options...** button.
- ▪ Choose **Linear regression** as the **Fit Method** on the resulting **Fit Line** sub-dialogue box.

This produces the scatterplot matrix shown in Display 4.13. The variable plotted on the *y*-axis in a given cell of the matrix is identified by reading along the corresponding row; the variable on the *x*-axis by reading along the corresponding column. The two scattergraphs in the first row of the scatter matrix display the dependent variable (minimum temperature)

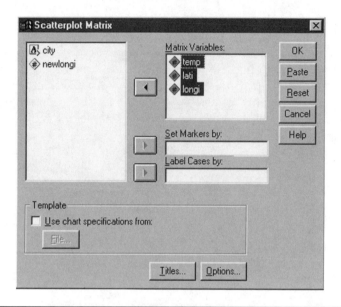

Display 4.12   Generating a scatterplot matrix.

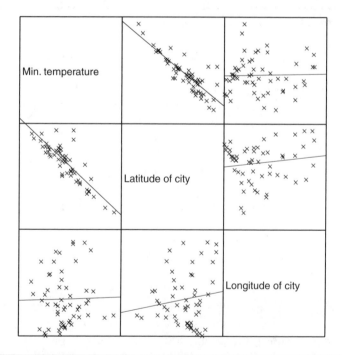

Display 4.13   Scatterplot matrix of temperature, latitude, and longitude.

on the *y*-axis and the explanatory variables latitude and longitude and the respective *x*-axes. The simple linear regressions lines show that the relationship between temperature and latitude is approximately linear, while the relationship between temperature and longitude appears to be more complex; perhaps taking an "S" shaped form that might need to be represented by a cubic term for longitude in the regression model. Consequently, we will consider the following model for the temperature data:

$$y_i = \beta_0 + \beta_1 x_{1i} + \beta_2 x_{2i} + \beta_3 x_{2i}^2 + \beta_3 x_{2i}^3 + \varepsilon_i \qquad (4.1)$$

where x_{1i} and x_{2i} are the latitude and longitude coordinates of the *i*th city, y_i the observed minimum January temperature, and ε_i an error term. Provided that longitude is constant, this equation models a linear relationship between temperature and latitude. When latitude is held constant, a cubic relationship between temperature and longitude is modeled. The multiple regression model now contains four explanatory variables x_1, x_2, x_2^2, and x_2^3. The linear regression coefficients β_0, β_1, β_2, β_3, and β_4 are, of course, unknown and, as always, will need to be estimated from the observed data.

To fit this regression model to the data, we first need to create each of the explanatory variables to be used in the model. To do this, we need the **Compute** command (see Chapter 1, Display 1.13) to create new variables c_lati, c_longi, c_longi2, and c_longi3. The variables c_lati and c_longi measure latitudes and longitudes as deviations from their sample means (38.97 and 90.964, respectively). c_longi2 and c_longi3 were then constructed by squaring or cubing new variable c_longi, respectively. Such *mean centering* (or subtraction of the sample mean) avoids computational problems (or *ill-conditioning*), a problem that can arise as a result of correlations between a positive predictor variable and its square or cube (see Rawlings, Patula, and Dickey, 1998). The regression model is then fitted by specifying temp as the dependent variable and c_lati, c_longi, c_longi2, and c_longi3 as the independent variables in the **Linear Regression** dialogue box (see Display 4.4).

The resulting SPSS output tables are shown in Display 4.14. The value of the multiple correlation coefficient, 0.973, indicates that the observed temperatures are fitted very well by those predicted by the estimated model. In terms of variability in minimum temperature accounted for by our fitted model, this amounts to a proportion of $R^2 = 0.964$, or 94.6%. The regression is clearly statistically significant ($F(4,51) = 223.9, p < 0.001$).

The "Coefficients" table provides the estimates of the regression coefficients. Here we estimate that minimum temperature reduces by 2.36°F for every unit increase in latitude when longitude remains unchanged. (In

Model Summary

Model	R	R Square	Adjusted R Square	Std. Error of the Estimate
1	.973[a]	.946	.942	3.225

a. Predictors: (Constant), C_LONGI3, C_LATI, C_LONGI2, C_LONGI

ANOVA[b]

Model		Sum of Squares	df	Mean Square	F	Sig.
1	Regression	9315.528	4	2328.882	223.908	.000[a]
	Residual	530.455	51	10.401		
	Total	9845.982	55			

a. Predictors: (Constant), C_LONGI3, C_LATI, C_LONGI2, C_LONGI
b. Dependent Variable: Average January minimum temperature

Coefficients[a]

Model		Unstandardized Coefficients		Standardized Coefficients	t	Sig.	95% Confidence Interval for B	
		B	Std. Error	Beta			Lower Bound	Upper Bound
1	(Constant)	23.130	.672		34.410	.000	21.780	24.479
	C_LATI	-2.358	.090	-.948	-26.204	.000	-2.538	-2.177
	C_LONGI	-.580	.068	-.649	-8.473	.000	-.717	-.442
	C_LONGI2	2.011E-03	.003	.040	.600	.551	-.005	.009
	C_LONGI3	1.296E-03	.000	.898	8.353	.000	.001	.002

a. Dependent Variable: Average January minimum temperature

Display 4.14 Output from multiple linear regression of minimum temperatures.

this example, it does not make sense to consider the other regression coefficients singly, since the effect of longitude is modeled jointly by the variables c_longi, c_longi2, and c_longi3.)

Having established that our model is highly predictive in terms of R^2, it is of interest to see how much of this predictive power is due to latitude and how much is due to longitude. As a result of some correlation between the latitude and longitude of American cities (Display 4.11), we cannot uniquely attribute part of the explained variability in temperatures to latitude or longitude. However, we can measure how much variability is explained by a model that contains latitude alone, and by how much this is increased after adding the effects of longitude to the model and vice versa. This requires us to fit a series of models as shown in Table 4.3.

In SPSS, "blocking" of the explanatory variables can be used to fit a series of models for a dependent variable derived by consecutively adding further terms. The first block contains the explanatory variables of the first model. The second block contains the explanatory variables that are to be added to the model in the second step of the model-fitting procedure. The third block contains the variables to be added in the third step and so on. When the explanatory variables are blocked, all regression output is generated repeatedly for each step of the modeling procedure.

Table 4.3 Stepwise Model Fitting for Temperature Data

Order of Model Terms	Step	Model Terms Added	Explanatory Variables in Model	Effect of Change in Model Fit Measures
Latitude first	1	Latitude term	c_lati	Latitude
	2	Longitude terms	c_lati, c_longi, c_longi2, c_longi3	Longitude not accounted for by latitude
Longitude first	1	Longitude terms	c_longi, c_longi2, c_longi3	Longitude
	2	Latitude term	c_longi, c_longi2, c_longi3, c_lati	Latitude not accounted for by longitude

We proceed to use the blocking facility to carry out the step-wise modeling of temperature as described in Table 4.3. Our first series of two models translates into block 1 containing only the latitude variable c_lati and block 2 containing the longitude variables c_longi, c_longi2, and c_longi3. The **Linear Regression** dialogue box informs SPSS about a blocking of the explanatory variables. When variables are to be added in blocks, only the variables of the first block are included under the **Independent(s)** list initially. Then, clicking the **Next** button causes SPSS to empty the **Independent(s)** list and to ask for the variables of the next block ("Block 2 of 2" is now displayed). The explanatory variables of the second block are then entered (Display 4.15). The **Next** button is repeatedly clicked until the last block has been defined. Setting **Method** to **Enter** ensures that variables are added in each step. (It is also possible to **Remove** the variables of a block from the model.) Here we are only interested in how R^2 changes in each step and hence selected **R squared change** in the **Statistics** sub-dialogue box (see Display 4.4).

The SPSS output generated by fitting our first series of models is shown in Display 4.16. Repeating the whole procedure, this time defining c_longi, c_longi2, and c_longi3 as block 1 and c_lati as block 2 leads to the output in Display 4.17. The output provides information on which variables are added in each step. It then produces a "Model Summary" table that contains the changes in R^2 and F-tests for the null hypotheses of no effects of the added variables. Display 4.16 shows that latitude alone explains 71.9% of the variance in temperatures and this contribution is statistically significant ($F(1,54) = 138.3$, $p < 0.001$). Adding the longitude terms to the model increases this further by 22.7%. The respective test shows that longitude significantly affects temperature after adjusting for latitude ($F(3,51) = 71.6$, $p < 0.001$). Reversing the order of model fitting makes

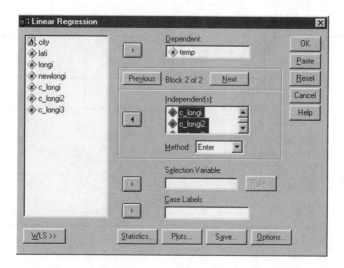

Display 4.15 Blocking explanatory variables of a regression model.

little difference. Longitude alone explains 22.1% of the variance in temperatures $(F(3,52) = 4.9, p = 0.004)$. Adding the latitude term accounts for a further 72.5%. The F-test for the latter $(F(1,51) = 686.6, p < 0.001)$ again shows that latitude significantly affects temperatures after adjusting for longitude. We can, therefore, be confident that both the linear effect of latitude and the third order polynomial of longitude are needed to model minimum January temperatures in the U.S.

Having arrived at a final multiple regression model for a data set, it is important to carry on and check the assumptions made in the modeling process. Only when the model appears adequate in light of the data should the fitted regression equation be interpreted. Three assumptions are made in a multiple regression modeling:

1. The errors have the same variance (homogeneity).
2. The errors arise from a normal distribution.
3. The relationship between each explanatory variable and the dependent variable is linear and the effects of several explanatory variables are additive.

The most useful approach for assessing these assumptions is to examine some form of *residual* from the fitted model along with some of the many other regression diagnostics now available (see Exercise 4.4.3). Residuals at their simplest are the difference between the observed and fitted values of the response variable. The following residual diagnostics are commonly used:

Variables Entered/Removed[b]

Model	Variables Entered	Variables Removed	Method
1	C_LATI[a]	.	Enter
2	C_LONGI, C_LONGI 2, C_LONGI3[a]	.	Enter

a. All requested variables entered.

b. Dependent Variable: Average January minimum temperature

Model Summary

		Change Statistics			
Model	R Square Change	F Change	df1	df2	Sig. F Change
1	.719[a]	138.283	1	54	.000
2	.227[b]	71.616	3	51	.000

a. Predictors: (Constant), C_LATI

b. Predictors: (Constant), C_LATI, C_LONGI, C_LONGI2, C_LONGI3

Display 4.16   Change in model fit when latitude term is fitted first.

Variables Entered/Removed[b]

Model	Variables Entered	Variables Removed	Method
1	C_LONGI 3, C_LONGI 2, C_LONGI[a]	.	Enter
2	C_LATI[a]	.	Enter

a. All requested variables entered.

b. Dependent Variable: Average January minimum temperature

Model Summary

		Change Statistics			
Model	R Square Change	F Change	df1	df2	Sig. F Change
1	.221[a]	4.911	3	52	.004
2	.725[b]	686.645	1	51	.000

a. Predictors: (Constant), C_LONGI3, C_LONGI2, C_LONGI

b. Predictors: (Constant), C_LONGI3, C_LONGI2, C_LONGI, C_LATI

Display 4.17   Change in model fit when longitude terms are fitted first.

1. *Residual plot*: This is a scatterplot of residuals against predicted values. The residuals should lie around zero with the degree of scatter not varying systematically with the size of predicted values. (More details can be found in Cook and Weisberg, 1982.)
2. *Any diagnostic plot for normality*: For example, box plot, QQ-plot, histogram (see Chapter 2).
3. *Partial plots*: A partial plot of an explanatory variable plots the raw residuals from a regression of the dependent variable on all other explanatory variables against the raw residuals from a regression of the explanatory variable in question on all other explanatory variables. The relationship between the variables should be approximately linear. (More details can again be found in Cook and Weisberg, 1982.)

The quickest way to produce a set of residual diagnostics is to check the **Plots...** button on the **Linear Regression dialogue** box (see Display 4.4) and fill in the fields of the **Plots** sub-dialogue box as illustrated in Display 4.18. SPSS provides a number of types of residuals and predicted values. (These are also listed under the **Save** sub-dialogue box, see Display 4.9.) *Unstandardized* or raw residuals are simply calculated as the difference between the observed values and those predicted by the model. By definition these have sample mean zero and are measured in units of the dependent variable, making them somewhat inconvenient to use in practice. To make them scale invariant, they are often standardized to units of estimated error standard deviation. SPSS refers to such scaled residuals as *standardized residuals* (*ZRESID). However, the variances of both, the raw residuals and the standardized residuals, vary with the values of the explanatory variables, and in that sense these residuals are not optimal for examining

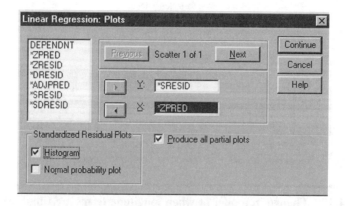

Display 4.18 Generating residual diagnostics.

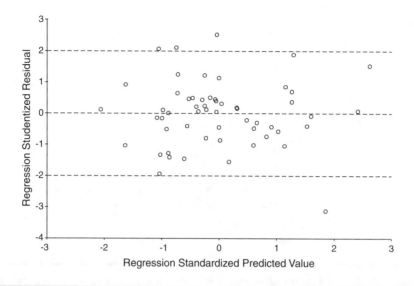

Display 4.19 Residual plot generated by commands in Display 4.18.

the error distribution (which has constant variance). *Studentized residuals* (*SRESID) are corrected for this effect of the explanatory variables and should, if the model is correct, arise from a standard normal distribution; it is these residuals that we shall use here.

We generate a residual plot by plotting *SRESID on the y-axis and *ZPRED (predicted values standardized to sample mean zero and standard deviation one) on the x-axis. This results in the scattergraph shown in Display 4.19. We further choose to examine the distributional shape of the residuals by means of a histogram and checked **Histogram**. (Unfortunately, SPSS only allows for plotting a histogram of the standardized residuals, Display 4.20.) Finally, we request partial plots by checking **Produce all partial plots**. This generates the four graphs shown in Display 4.21.

The residual plot was edited somewhat for appearance (Display 4.19). In particular, three horizontal reference lines were included to examine outliers. SPSS enables inclusion of horizontal lines by using the commands

Chart – Reference Line... – Y Scale

in the **Chart Editor**. This opens the **Scale Axis Reference Line** dialogue box where the values at which the reference lines are to cross the y-axis can be specified. Here we chose to insert three horizontal reference lines at y-levels of 2, 0, and –2, respectively. Under the model assumptions, 95% of the Studentized residuals are expected to be within the range from –2 to 2. We find that 4 out of 56 cities (7.1%) receive a residual with absolute value more than 2, which is perfectly reasonable under the model assumptions.

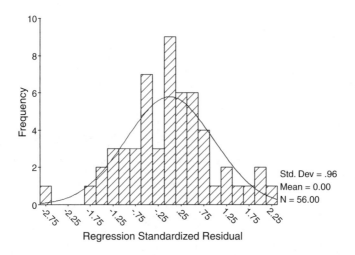

Display 4.20 Histogram of standardized residuals generated by commands in Display 4.18.

Examining the residual plot, it appears that the homogeneity of variance assumption is not violated since the residuals scatter randomly around the zero line and the degree of scatter appears constant across the range of predicted values. The histogram of the residuals is also consistent with the assumption of normality (Display 4.20.)

The partial plots in Display 4.21 can be made more useful by including the regression line from a simple linear regression of the variable on the *y*-axis (the temperature residual) and on the variable on the *x*-axis (the explanatory variable residual). This helps to assess whether there is linear relationship between the two variables. The partial plot for latitude shows a clear linear relationship (Display 4.21a). As a result of modeling the effect of longitude by three explanatory variables, three partial plots have been produced for longitude (Display 4.21b–d). All three plots show approximately linear relationships. We, therefore, conclude that the relationship between temperature and latitude and longitude has been modeled reasonably well by our multiple regression equation.

Having confirmed that all our model assumptions seem reasonable in the light of the data, we finally return to our objective for developing the multiple regression model, namely, using it to predict January minimum temperatures of U.S. locations not in the original sample. The "Coefficients" table in Display 4.14 provided the estimates of our regression parameters, giving $\hat{\beta}_0 = 23.31$, $\hat{\beta}_1 = -2.358$, $\hat{\beta}_2 = -0.58$, $\hat{\beta}_3 = 0.002011$, and $\hat{\beta}_4 = 0.001296$. (Note that SPSS displayed the last two coefficients in scientific notation to save space.) Our fitted regression equation is therefore

$$y = 23.13 - 2.358(x_1 - 38.97) - 0.58(x_2 - 90.964) +$$
$$0.002011(x_2 - 90.964)^2 + 0.001296(x_2 - 90.964)^3$$

(4.2)

where y is the January minimum temperature and x_1 and x_2 are the latitude and longitude of the location, respectively. Provided we know the latitude (x_1) and longitude (x_2) coordinates of a place, we can use this equation to predict its January minimum temperature (y). We can get SPSS to do the necessary calculation and also provide a measure of the precision of the prediction.

For illustrative purposes, we shall predict long-term January minimum temperatures at an average U.S. latitude (the sample mean = 38.97) and across a range of longitudes (125, 120, 115, ..., 80). To obtain predictions at these values of the explanatory variables, we need to extend the **Data View** spreadsheet so that it includes the new values of the explanatory variables as shown in Display 4.22. We have labeled the new locations "new location 1" to "...10" and inserted the latitude and longitude values for which we want to predict. However, since our regression operates on the centered variables **c_lati**, **c_longi**, **c_longi2**, and **c_longi3**, these also have to be filled in. The inverse longitude variable **newlongi** is also specified since we later want to use it for plotting. Obviously, the temperature variable **temp** remains blank since it is to be predicted.

Once the spreadsheet has been extended, we can run the regression fitting as before, except that this time we check **Unstandardized Predicted Values** and **Mean Prediction Intervals** in the **Save** sub-dialogue box (see Display 4.9). This provides three new variables on the right-hand side of the original spreadsheet: the predicted value at each location including the ten new locations (**pre_1**), and the lower (**lmci_1**) and upper limit (**umci_1**) of a 95% confidence interval for the long-term January minimum temperature. (Note that we could have also chosen a confidence interval for the January minimum temperature over a single 30-year period by checking **Individual Prediction Intervals** in the **Save** sub-dialogue box.)

Finally, we need to display the obtained predictions and confidence intervals. To do this, we first restrict the spreadsheet to the new locations by using the following steps:

- Use the **Select Cases...** command from the **Data** drop-down menu (see Chapter 1, Display 1.11).
- Check **If condition is satisfied** in the resulting dialog and click the button labeled **If....**
- Insert the condition "$casenum>56" in the resulting **Select Cases: If** sub-dialogue box. ($casenum is a system variable and allows us to access the row numbers.)

a) Latitude

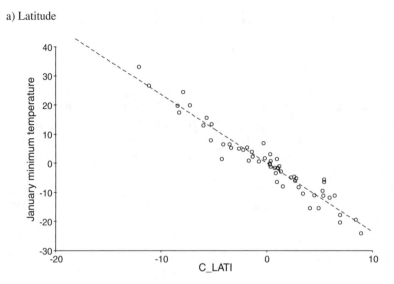

b) Longitude

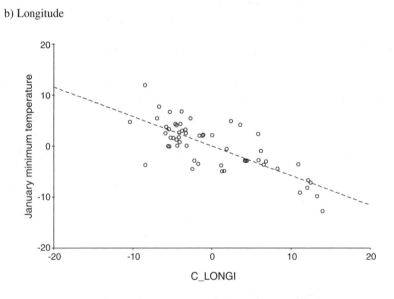

Display 4.21   Partial plots generated by commands in Display 4.18.

c) Longitude squared

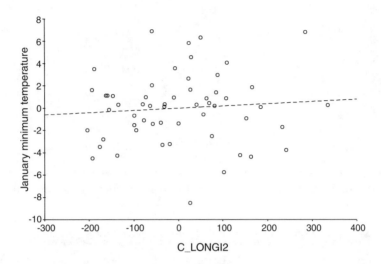

d) Longitude cubed

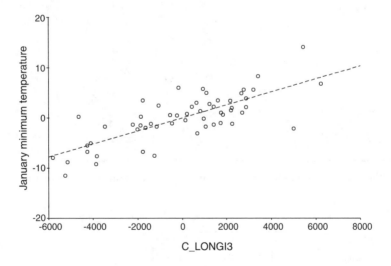

Display 4.21 (continued)

	city	temp	lati	longi	newlongi	c_lati	c_longi	c_longi2	c_longi3
52	Seattle, WA	33	48.10	122.50	-122.50	9.13	31.54	994.52	31363.16
53	Spokane, WA	19	48.10	117.90	-117.90	9.13	26.94	725.55	19543.36
54	Madison, WI	9	43.40	90.20	-90.20	4.43	-.76	.58	-.45
55	Milwaukee, WI	13	43.30	88.10	-88.10	4.33	-2.86	8.20	-23.49
56	Cheyenne, WY	14	41.20	104.90	-104.90	2.23	13.94	194.21	2706.54
57	new location 1	.	38.97	125.00	-125.00	.00	34.04	1158.45	39428.98
58	new location 2	.	38.97	120.00	-120.00	.00	29.04	843.09	24479.94
59	new location 3	.	38.97	115.00	-115.00	.00	24.04	577.73	13886.30
60	new location 4	.	38.97	110.00	-110.00	.00	19.04	362.37	6898.06
61	new location 5	.	38.97	105.00	-105.00	.00	14.04	197.01	2765.22
62	new location 6	.	38.97	100.00	-100.00	.00	9.04	81.65	737.78
63	new location 7	.	38.97	95.00	-95.00	.00	4.04	16.29	65.74
64	new location 8	.	38.97	90.00	-90.00	.00	-.96	.93	-.90
65	new location 9	.	38.97	85.00	-85.00	.00	-5.96	35.57	-212.14
66	new location 10	.	38.97	80.00	-80.00	.00	-10.96	120.21	-1317.97

Display 4.22   Extending a spreadsheet to make predictions at new values of explanatory variables.

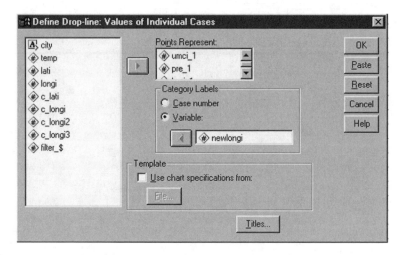

Display 4.23   Graphing a series of precalculated confidence intervals.

Executing these commands results in rows 1 to 56 of the **Data View** spreadsheet being crossed out and ignored in all further SPSS procedures. We then proceed to construct a graph that displays the predicted temperatures and CIs for each longitude as follows:

- Choose the commands **Graphs – Line....**
- Check **Drop-line** and **Values of individual cases** in the resulting **Line Charts** dialogue box.
- Fill the next dialogue box in as indicated in Display 4.23.

This results in the graph shown in Display 4.24 (again after some editing for better presentation). The predicted temperature values and CIs

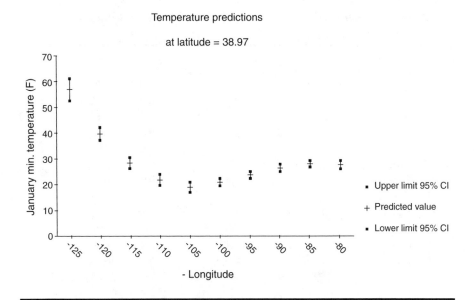

Display 4.24 Predicted long-term January minimum temperatures across the U.S.

are plotted against (inverse) longitude. For a latitude of 38.97, we can see that long-term January minimum temperatures are predicted to be higher on the west coast than on the east coast and that the lowest temperatures are predicted for inland U.S. The widths of the confidence intervals show that temperatures at Eastern locations can be predicted more precisely than at Western locations, reflecting the denser city coverage in the east of the U.S. (see Display 4.11).

Finally, having developed a prediction equation, we would like to conclude with a warning. Regression should not be used to carry out predictions outside the range of the explanatory variables; for example, here outside the area of the U.S. In addition, predictions carried out at levels of unobserved variables not comparable to the observed data can result in very misleading predictions. Here, for example, predictions at high altitudes are bound to be too high, given that temperature is known to decease with increasing altitude. (For more details on the pitfall of regression predictions, see Harrell, 2000).

4.4 Exercises

4.4.1 Air Pollution in the U.S.

The data in Table 4.4 relate to air pollution in 41 U.S. cities. Use multiple linear regression to investigate determinants of pollution. Air pollution

Table 4.4   Sulfur Dioxide and Indicators of Climatic Conditions and Human Ecology

City	SO₂	Temp.	Manuf.	Pop	Wind	Precip.	Days
Phoenix	10	70.3	213	582	6.0	7.05	36
Little Rock	13	61.0	91	132	8.2	48.52	100
San Francisco	12	56.7	453	716	8.7	20.66	67
Denver	17	51.9	454	515	9.0	12.95	86
Hartford	56	49.1	412	158	9.0	43.37	127
Wilmington	36	54.0	80	80	9.0	40.25	114
Washington	29	57.3	434	757	9.3	38.89	111
Jacksonville	14	68.4	136	529	8.8	54.47	116
Miami	10	75.5	207	335	9.0	59.80	128
Atlanta	24	61.5	368	497	9.1	48.34	115
Chicago	110	50.6	3344	3369	10.4	34.44	122
Indianapolis	28	52.3	361	746	9.7	38.74	121
Des Moines	17	49.0	104	201	11.2	30.85	103
Wichita	8	56.6	125	277	12.7	30.58	82
Louisville	30	55.6	291	593	8.3	43.11	123
New Orleans	9	68.3	204	361	8.4	56.77	113
Baltimore	47	55.0	625	905	9.6	41.31	111
Detroit	35	49.9	1064	1513	10.1	30.96	129
Minneapolis	29	43.5	699	744	10.6	25.94	137
Kansas	14	54.5	381	507	10.0	37.00	99
St. Louis	56	55.9	775	622	9.5	35.89	105
Omaha	14	51.5	181	347	10.9	30.18	98
Albuquerque	11	56.8	46	244	8.9	7.77	58
Albany	46	47.6	44	116	8.8	33.36	135
Buffalo	11	47.1	391	463	12.4	36.11	166
Cincinnati	23	54.0	462	453	7.1	39.04	132
Cleveland	65	49.7	1007	751	10.9	34.99	155
Columbus	26	51.5	266	540	8.6	37.01	134
Philadelphia	69	54.6	1692	1950	9.6	39.93	115
Pittsburgh	61	50.4	347	520	9.4	36.22	147
Providence	94	50.0	343	179	10.6	42.75	125
Memphis	10	61.6	337	624	9.2	49.10	105
Nashville	18	59.4	275	448	7.9	46.00	119
Dallas	9	66.2	641	844	10.9	35.94	78
Houston	10	68.9	721	1233	10.8	48.19	103
Salt Lake City	28	51.0	137	176	8.7	15.17	89
Norfolk	31	59.3	96	308	10.6	44.68	116
Richmond	26	57.8	197	299	7.6	42.59	115
Seattle	29	51.1	379	531	9.4	38.79	164

Table 4.4 (continued) Sulfur Dioxide and Indicators of Climatic Conditions and Human Ecology

City	SO₂	Temp.	Manuf.	Pop	Wind	Precip.	Days
Charleston	31	55.2	35	71	6.5	40.75	148
Milwaukee	16	45.7	569	717	11.8	29.07	123

SO₂: Sulfur dioxide content of air in micrograms per cubic meter
TEMP: Average annual temperature °F
MANUF: Number of manufacturing enterprises employing 20 or more workers
POP: Population size (1970 census) in thousands
WIND: Average wind speed in miles per hour
PRECIP: Average annual precipitation in inches
DAYS: Average number of days with precipitation per year

Source: Hand et al., 1994.

(SO₂) is measured by the annual mean concentration of sulfur dioxide, in micrograms per cubic meter. The other six variables are indicators of climate conditions and human ecology.

4.4.2 Body Fat

The data in Table 4.5 come from a study investigating a new method of measuring body composition and gives the body fat percentage (% fat), age, and sex for 25 normal adults between 23 and 61 years.

1. Construct a scatterplot of % fat and age within subgroups defined by gender. Does it appear as if the relationship is different for males and females? (Hint: Use the commands Graphs – Scatter... – Simple – Set Markers by: sex)
2. Model the relationship between % fat and age and gender.

4.4.3 More on Cleaning Cars: Influence Diagnostics

A variety of diagnostic measures for identifying cases with great influence on the results of a regression have been developed in the past decade or so. Three that are commonly used are:

■ *DfBeta residual*: Measures the difference in the regression coefficient estimate of an explanatory variable when a particular observation is included and when it is excluded in the calculation.
■ *Standardized DfBeta residual*: As above but now scaled to have unit variance, a value exceeding $2/\sqrt{n}$ is often viewed as indicating

Table 4.5   Body Fat Content and Age of Adults

Age (years)	% Fat	Sex (0 = male, 1 = female)
23	9.5	0
23	27.9	1
27	7.8	0
27	17.8	0
39	31.4	1
41	25.9	1
45	27.4	0
49	25.2	1
50	31.1	1
53	34.7	1
53	42.0	1
54	42.0	1
54	29.1	1
56	32.5	1
57	30.3	1
57	21.0	0
58	33.0	1
58	33.8	1
60	41.1	1
61	34.5	1

Source: Hand et al., 1994.

that the associated observation has undue influence of the estimate of the particular regression coefficient (Miles and Shevlin, 2001).

■ *Cook's distance statistic:* calculated for each observation, this gives a measure of the combined change in the estimates of the regression coefficients that would result from excluding the observation (Cook, 1977).

Investigate the use of these diagnostics on the car cleaning data by looking under the "Distances" and "Influence Statistics" sections of the **Linear Regression: Save** sub-dialogue box.

Chapter 5

Analysis of Variance I: One-Way Designs; Fecundity of Fruit Flies, Finger Tapping, and Female Social Skills

5.1 Description of Data

Three data sets will be analyzed in this chapter; each of them arises from what is generally known as a *one-way design* in which one (or more than one) continuous outcome variable(s) is observed on a sample of individuals grouped according to a single categorical variable or factor. The question addressed in a one-way design is: Do the populations that give rise to the different levels of the factor variable have different mean values on the outcome variable?

The first data set, shown in Table 5.1, arises from a study of the fecundity of the fruit fly *Drosophila melanogaster*. For 25 female flies from each of three strains, per diem fecundity (number of eggs laid per female per day for the first 14 days of life) was recorded. The genetic lines labeled RS and SS were selectively bred for resistance and for susceptibility to the pesticide, DDT, and the NS line is a nonselected control strain. In this

Table 5.1   Fecundity of Fruit Flies

Resistant (RS)	Susceptible (SS)	Nonselected (NS)
12.8	38.4	35.4
21.6	32.9	27.4
14.8	48.5	19.3
23.1	20.9	41.8
34.6	11.6	20.3
19.7	22.3	37.6
22.6	30.2	36.9
29.6	33.4	37.3
16.4	26.7	28.2
20.3	39.0	23.4
29.3	12.8	33.7
14.9	14.6	29.2
27.3	12.2	41.7
22.4	23.1	22.6
27.5	29.4	40.4
20.3	16.0	34.4
38.7	20.1	30.4
26.4	23.3	14.9
23.7	22.9	51.8
26.1	22.5	33.8
29.5	15.1	37.9
38.6	31.0	29.5
44.4	16.9	42.4
23.2	16.1	36.6
23.6	10.8	47.4

Source: Hand et al., 1994.

experiment, the effect of the factor "Genetic strain" on fecundity is of interest.

The second data set, shown in Table 5.2, comes from an investigation into the effect of the stimulant caffeine on the performance of a simple task. Forty male students were trained in finger tapping. They were then divided at random into 4 groups of 10, and the groups received different doses of caffeine (0, 100, 200, and 300 ml). Two hours after treatment, each student was required to carry out finger tapping and the number of taps per minute was recorded. In this *dose–response experiment*, the main question of interest is how the amount of finger tapping changes with the increasing amount of caffeine given.

Table 5.2 Finger Tapping and Caffeine Consumption

0 ml	100 ml	200 ml	300 ml
242	248	246	248
245	246	248	250
244	245	250	251
248	247	252	251
247	248	248	248
248	250	250	251
242	247	252	252
244	246	248	249
246	243	245	253
242	244	250	251

Source: Everitt, 2001.

The third data set we shall deal with in this chapter is shown in Table 5.3. These data come from a study concerned with improving the social skills of college females and reducing their anxiety in heterosexual encounters (see Novince, 1977). Women who agreed to take part in the investigation were randomly allocated to one of three training regimes: a control group, a behavioral rehearsal group, and a behavioral rehearsal and plus cognitive restructuring group. The values on the following four dependent variables were recorded for each subject in the study:

- Anxiety — physiological anxiey in a series of heterosexual encounters
- Measure of social skills in social interactions
- Appropriateness
- Assertiveness

These four variables share a common conceptual meaning; they might all be regarded as measures of the underlying concept "social confidence." Consequently, here the question of primary interest is: Do the four dependent variables as a whole suggest any difference between the training regimes?

5.2 Analysis of Variance

The phrase "analysis of variance" was coined by arguably the most famous statistician of the twentieth century, Sir Ronald Aylmer Fisher, who defined it as "the separation of variance ascribable to one group of causes from

Table 5.3 Female Social Skills

Anxiety	Social Skills	Appropriateness	Assertiveness	Group[a]
5	3	3	3	1
5	4	4	3	1
4	5	4	4	1
4	5	5	4	1
3	5	5	5	1
4	5	4	4	1
4	5	5	5	1
4	4	4	4	1
5	4	4	3	1
5	4	4	3	1
4	4	4	4	1
6	2	1	1	2
6	2	2	2	2
5	2	3	3	2
6	2	2	2	2
4	4	4	4	2
7	1	1	1	2
5	4	3	3	2
5	2	3	3	2
5	3	3	3	2
5	4	3	3	2
6	2	3	3	2
4	4	4	4	3
4	3	4	3	3
4	4	4	4	3
4	5	5	5	3
4	5	5	5	3
4	4	4	4	3
4	5	4	4	3
4	6	6	5	3
4	4	4	4	3
5	3	3	3	3
4	4	4	4	3

a Group codes: 1 = behavioral rehearsal, 2 = control, 3 = behavioral rehearsal and cognitive restructuring.

the variance ascribable to the other groups". Stated another way, the analysis of variance (ANOVA) is a partitioning of the total variance in a set of data into a number of component parts, so that the relative contributions of identifiable sources of variation to the total variation in measured responses can be determined. From this partition, suitable *F*-tests can be derived that allow differences between sets of means to be assessed. Details of the analysis of variance for a one-way design with a single response variable are given in Box 5.1

When a set of dependent variables is to be compared in a one-way design, the multivariate analogue of the one-way analysis of variance described in Box 5.1 is used. The hypothesis tested is that the set of variable means (the *mean vector*) is the same across groups. Details are given in Everitt (2001b). Unfortunately, in the multivariate situation (unless there are only two groups to be compared), there is no single test statistic that can be derived for detecting all types of possible departures from the null hypothesis of the equality of the group mean vectors. A number of different test statistics have been proposed, as we shall see when we come to analyze the data in Table 5.3. All such test statistics can be transformed into *F*-statistics to enable p-values to be calculated. (All the test statistics are equivalent in the two-group situation.)

Multivariate analysis of variance (MANOVA) assumes multivariate normality of the variables in each factor level and a common covariance matrix (see Everitt, 2001b, for further details).

Box 5.1 One-Way Analysis of Variance

- The model assumed for the observations from a one-way design is

$$y_{ij} = \mu_i + \varepsilon_{ij}$$

  where y_{ij} represents the *j*th observation in the *i*th group, and the ε_{ij} represent random error terms, assumed to be from a normal distribution with mean zero and variance σ^2.
- The null hypothesis of the equality of population means can now be written as

$$H_0: \mu_1 = \mu_2 = \cdots = \mu_k = \mu$$

  leading to a new model for the observations, namely

$$y_{ij} = \mu + \varepsilon_{ij}$$

■ There are some advantages (and, unfortunately, some disadvantages) in reformulating the model slightly, by modeling the mean value for a particular population as the sum of the overall mean value of the response plus a specific population or group effect. This leads to a *linear model* of the form

$$y_{ij} = \mu + \alpha_i + \varepsilon_{ij}$$

where μ represents the overall mean of the response variable, α_i is the effect on an observation of being in the ith group ($i = 1, 2, ..., k$), and again ε_{ij} is a random error term, assumed to be from a normal distribution with mean zero and variance σ^2.

■ When written in this way, the model uses $k + 1$ parameters ($\mu, \alpha_1, \alpha_2, ..., \alpha_k$) to describe only k group means. In technical terms, the model is said to be *overparameterized*, which causes problems because it is impossible to find unique estimates for each parameter — it is a bit like trying to solve simultaneous equations when there are fewer equations than unknowns. The following constraint is generally applied to the parameters to overcome the problem (see Maxwell and Delaney, 1990, for details):

$$\sum_{i=1}^{k} \alpha_i = 0$$

■ If this model is assumed, the hypothesis of the equality of population means can be rewritten in terms of the parameters α_i as

$$H_0: \alpha_1 = \alpha_2 = \cdots = \alpha_k = 0,$$

so that under H_0 the model assumed for the observations is

$$y_{ij} = \mu + \varepsilon_{ij}$$

as before.

■ The necessary terms for the *F*-test are usually arranged in an *analysis of variance table* as follows (*N* is the total number of observations).

Source of Variation	DF	SS	MS	MSR (F)
Between groups	$k - 1$	BGSS	BGSS/$(k - 1)$	MSBG/MSWG
Within groups (error)	$N - k$	WGSS	WGSS/$(N - k)$	
Total	$N - 1$			

Here, DF is degrees of freedom, SS is sum of squares, MS is mean square, BGSS is between groups sum of squares, and WGSS is within group sum of squares.

■ If H_0 is true and the assumptions listed below are valid, the mean square ratio (MSR) has an *F*-distribution with $k - 1$ and $N - k$ degrees of freedom.

■ The data collected from a one-way design have to satisfy the following assumptions to make the *F*-test involved strictly valid:

1. The observations in each group come from a normal distribution.
2. The population variances of each group are the same.
3. The observations are independent of one another.

5.3 Analysis Using SPSS

5.3.1 Fecundity of Fruit Flies

Conceptually, the data in Table 5.1 represent a one-way design, with factor "strain" (three levels: "RS," "SS," and "NS") and 25 replicate measures of per diem fecundity. To convey the one-way structure of the experiment, a **Data View** spreadsheet has to be set up as shown in Display 5.1. The rows of the spreadsheet correspond to fruit flies, and for each fly we have recorded its fecundity (variable **fecundit**) and its genetic line (variable **strain**).

Box plots provide a useful tool for an initial examination of grouped continuous data, and here we produce box plots of the fecundities of each strain of fruit fly. We have already seen in Chapter 2 how to generate box plots within groups using

Analyze – Descriptive statistics – Explore

	fecundit	strain
1	12.8	1
2	21.6	1
3	14.8	1
4	23.1	1
5	34.6	1
6	19.7	1
7	22.6	1
8	29.6	1
9	16.4	1
10	20.3	1
11	29.3	1
12	14.9	1
13	27.3	1
14	22.4	1
15	27.5	1
16	20.3	1
17	38.7	1
18	26.4	1
19	23.7	1
20	26.1	1
21	29.5	1
22	38.6	1
23	44.4	1
24	23.2	1
25	23.6	1
26	38.4	2
27	32.9	2
28	48.5	2
29	20.9	2

Display 5.1   SPSS spreadsheet containing data from Table 5.1.

To illustrate a further feature of SPSS we will, in this case, create a set of box plots from the graph menu by using the following steps:

■ Graphs – Boxplot...
■ Keep the default settings of the resulting **Boxplot** dialogue box (**Simple** box plot and display of **Summaries for groups of cases**).
■ Define **Variable** to be **fecundit** and **Category Axis** to be **strain** in the next **Define Simple Boxplot: Summaries for Groups of Cases** dialogue box.

The resulting set of box plots is shown in Display 5.2. This plot is useful in an informal assessment of both the homogeneity and normality assumptions of ANOVA. Here, the heights of the boxes, which measure

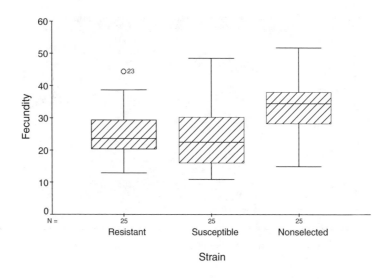

Display 5.2   Box plots of per diem fecundities for three strains of fruit fly.

the inter-quartile ranges, appear approximately constant across strain. Consequently, the homogeneity of variance assumption seems reasonable. And, except perhaps for the susceptible strain, the distribution within the fruit fly groups appears reasonably symmetric, suggesting that the normality assumption is also acceptable. (A better way of assessing the normality assumption is by using residuals, an issue we shall return to in the next chapter.)

To employ the one-way ANOVA routine in SPSS and also derive some useful summary statistics for each group of observations, we need to choose

Analyze – Compare Means – One-Way ANOVA...

from the menu bar and then fill in the **One-Way ANOVA** dialogue box as shown in Display 5.3. Within-group summary statistics are requested by checking **Descriptive** in the **Options...** sub-dialogue box. A more formal test of the homogeneity assumption can be obtained by checking **Homogeneity of variance test**. The resulting output is shown in Display 5.4.

The "Descriptives" table shows that the nonselected strain has the highest mean fecundity (33.4 eggs per female per day for the first 14 days of life), followed by the resistant strain (25.3 eggs), and the susceptible strain (23.6 eggs). The table also includes 95% confidence intervals for the mean fecundity in each group. (Note that these confidence intervals are calculated using only the fecundity observations of the relevant group and so do not depend upon the homogeneity of variance assumption.)

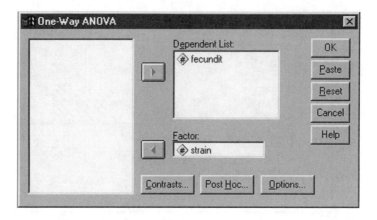

Display 5.3 Defining a one-way design.

Descriptives

FECUNDIT

	N	Mean	Std. Deviation	Std. Error	95% Confidence Interval for Mean Lower Bound	Upper Bound	Minimum	Maximum
Resistant	25	25.256	7.7724	1.5545	22.048	28.464	12.8	44.4
Susceptible	25	23.628	9.7685	1.9537	19.596	27.660	10.8	48.5
Nonselected	25	33.372	8.9420	1.7884	29.681	37.063	14.9	51.8
Total	75	27.419	9.7407	1.1248	25.178	29.660	10.8	51.8

Test of Homogeneity of Variances

FECUNDIT

Levene Statistic	df1	df2	Sig.
.748	2	72	.477

Display 5.4 Descriptive statistics and Levene's test for the one-way design.

The "Test of Homogeneity of Variances" table provides Levene's test for testing the null hypothesis that the within-group variances are constant across groups (Levene, 1960; see also the independent samples t-test in Chapter 2). In this instance, consistent with our examination of the box plots, the formal test does not find any evidence for a departure from the homogeneity assumption ($p = 0.48$).

We can now move on to formally assess the effect of strain on per diem fecundity. The one-way ANOVA table, including the required F-test (see Box 5.1), is automatically produced by the commands in Display 5.3 and is shown in Display 5.5. Here we find a significant effect of strain on fecundity ($F(2,72) = 8.7$, $p < 0.001$).

When a significant result has been obtained from an overall F-test, investigators often wish to undertake further tests to determine which

ANOVA

FECUNDIT

	Sum of Squares	df	Mean Square	F	Sig.
Between Groups	1362.211	2	681.106	8.666	.000
Within Groups	5659.022	72	78.598		
Total	7021.234	74			

Display 5.5   ANOVA table for fecundity data.

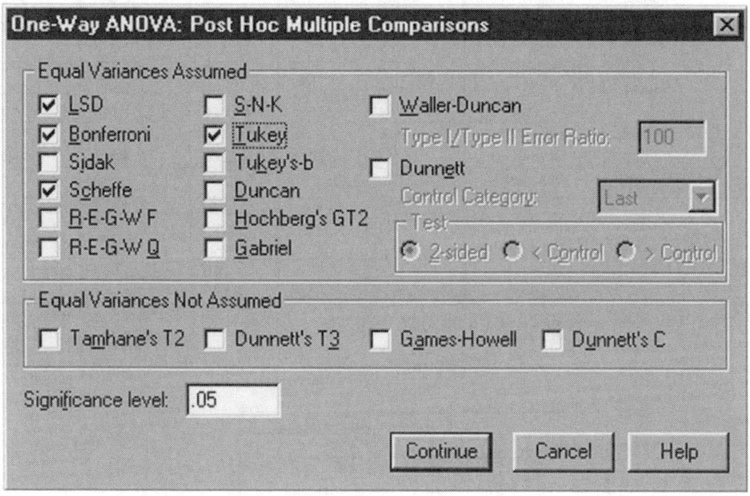

Display 5.6   Generating multiple comparisons.

particular group means differ. A number of procedures generically known as *multiple comparison techniques* can be employed for this purpose (for more details, see Everitt, 2001b). Such procedures aim to retain the nominal significance level at the required value when undertaking multiple tests In other words, such tests protect against claiming too many significant results (too many false positives). SPSS refers to these procedures as "*post hoc* multiple comparisons" since they should only be carried out once an overall effect of the grouping factor has been established. They can be requested via the **Post Hoc...** button on the One-Way ANOVA dialogue box (see Display 5.3). This opens the **Post Hoc Multiple Comparisons** sub-dialogue box shown in Display 5.6, where, for illustrative purposes, we have requested *Fisher's least significant differences* (LSD), *Bonferroni adjusted comparisons, Scheffe's multiple comparisons,* and *Tukey's honestly significant*

Multiple Comparisons

Dependent Variable: FECUNDIT

	(I) STRAIN	(J) STRAIN	Mean Difference (I-J)	Std. Error	Sig.	95% Confidence Interval Lower Bound	95% Confidence Interval Upper Bound
Tukey HSD	Resistant	Susceptible	1.628	2.5075	.793	-4.373	7.629
		Nonselected	-8.116*	2.5075	.005	-14.117	-2.115
	Susceptible	Resistant	-1.628	2.5075	.793	-7.629	4.373
		Nonselected	-9.744*	2.5075	.001	-15.745	-3.743
	Nonselected	Resistant	8.116*	2.5075	.005	2.115	14.117
		Susceptible	9.744*	2.5075	.001	3.743	15.745
Scheffe	Resistant	Susceptible	1.628	2.5075	.810	-4.640	7.896
		Nonselected	-8.116*	2.5075	.008	-14.384	-1.848
	Susceptible	Resistant	-1.628	2.5075	.810	-7.896	4.640
		Nonselected	-9.744*	2.5075	.001	-16.012	-3.476
	Nonselected	Resistant	8.116*	2.5075	.008	1.848	14.384
		Susceptible	9.744*	2.5075	.001	3.476	16.012
LSD	Resistant	Susceptible	1.628	2.5075	.518	-3.371	6.627
		Nonselected	-8.116*	2.5075	.002	-13.115	-3.117
	Susceptible	Resistant	-1.628	2.5075	.518	-6.627	3.371
		Nonselected	-9.744*	2.5075	.000	-14.743	-4.745
	Nonselected	Resistant	8.116*	2.5075	.002	3.117	13.115
		Susceptible	9.744*	2.5075	.000	4.745	14.743
Bonferroni	Resistant	Susceptible	1.628	2.5075	1.000	-4.519	7.775
		Nonselected	-8.116*	2.5075	.005	-14.263	-1.969
	Susceptible	Resistant	-1.628	2.5075	1.000	-7.775	4.519
		Nonselected	-9.744*	2.5075	.001	-15.891	-3.597
	Nonselected	Resistant	8.116*	2.5075	.005	1.969	14.263
		Susceptible	9.744*	2.5075	.001	3.597	15.891

*. The mean difference is significant at the .05 level.

FECUNDIT

Tukey HSD[a]

STRAIN	N	Subset for alpha = .05 — 1	Subset for alpha = .05 — 2
Susceptible	25	23.628	
Resistant	25	25.256	
Nonselected	25		33.372
Sig.		.793	1.000

Means for groups in homogeneous subsets are displayed.
a. Uses Harmonic Mean Sample Size = 25.000.

Display 5.7 Multiple comparisons output for fecundity data.

differences (HSD). (For details of these and the remaining multiple comparison techniques, see Everitt, 2001b, and Howell, 2002.)

The resulting multiple comparison output for the fruit fly data is shown in Display 5.7. For each comparison type and pair of groups, the "Multiple Comparisons" table provides an estimate of the difference in means, the standard error of that estimator, the p-value from a statistical test of zero group difference (under heading "Sig"), and a confidence interval for the mean difference. Each pair of groups appears twice in the table, once as a positive and once as a negative difference.

The LSD comparisons appear to achieve the lowest p-values. The reason for this is that the LSD method does *not* adjust for multiple comparisons. It is simply a series of *t*-tests using the common within-group standard deviation (and should, in our view, not appear under the multiple comparison heading). Thus, if the LSD method is to be used, we still need to adjust both the test and the confidence interval to correct for multiple testing. We can achieve this by employing the *Bonferroni correction,* which maintains the nominal significance level, α, by judging a group difference statistically significant when its p-value is below α/k, where k is the number of post hoc tests to be carried out (here three tests). According to the Bonferroni corrected LSD comparisons, both the SS and the RS strain differ significantly from the nonselected (NS) strain since the respective p-values are below $0.017 \approx 0.05/3$.

In fact, SPSS tries to carry out the Bonferroni correction automatically when **Bonferroni** is ticked in the **Post Hoc Multiple Comparisons** sub-dialogue box. To allow comparison against the nominal significance level, SPSS simply inflates the p-values of the *t*-tests by the factor k. For our data example, this means that the LSD p-values have been multiplied by the factor 3 (and set to 1 should the inflated value exceed 1) to derive "Bonferroni p-values". Bonferroni corrected confidence intervals are constructed by generating single confidence intervals at the $(1 - \alpha/k) \times 100\%$ confidence level, here the 98.3% confidence level.

The results from the Scheffe and Tukey procedures are, for this example, the same as given by the LSD and Bonferroni approach.

For some multiple comparison procedures, including Tukey's and Scheffe's, SPSS identifies subsets of groups whose means do not differ from one another. The second table in Display 5.7 shows these "homogeneous subsets" according to Tukey's comparisons. The RS and SS strain are allocated to the same subset of strains while the NS strain defines its own strain subset.

5.3.2 *Finger Tapping and Caffeine Consumption*

The one-way design of the finger tapping study was translated into a **Data View** spreadsheet as described for the fruit fly example. The spreadsheet contains a finger tapping variable (**tap**) and a caffeine dose factor (**dose**). In this case, we will use an error bar graph to get a first impression of the data. This requires the instruction

Graph – Error Bar…

and then the following steps:

- Click Simple in the resulting Error Bar dialogue box.
- Select Summaries of groups of cases.
- In the next Define Simple Error Bar: Summaries for Groups of Cases dialogue box, set Variable to taps.
- Set Category Axis to dose.
- Set Bars Represent to Standard deviation.
- Set Multiplier to 2.

The resulting graph is shown in Display 5.8. Since we have selected bars to be plotted at two standard deviations, approximately 95% of the finger tapping observations would be expected to lie between the upper and lower bars if the normality assumption is reasonable. There is a strong suggestion in Display 5.8 that the average finger tapping frequency increases with increasing amount of caffeine given.

Conceptually, the caffeine study is different from the fruit fly study in that a specific experimental hypothesis about the caffeine groups is of interest *a priori,* namely, that mean finger tapping frequency in the underlying population changes gradually with increasing level of the dose factor. In other words, we want to assess whether there is a trend in mean finger tapping frequency over dose. Such trends can be assessed relatively simply by using what are known as *orthogonal polynomials* (see Everitt, 2001b). Essentially, each such polynomial is a linear combination of the means in each factor level; different coefficients define combinations of means that measure the size of trends of different types (linear, quadratic, etc.). How the coefficients are chosen is not of great importance here, but we should

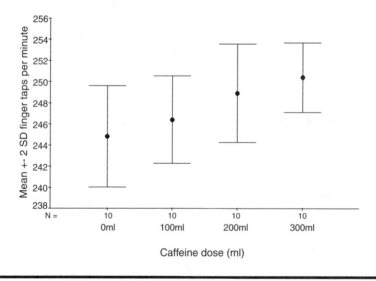

Display 5.8   Error bar graph for the finger tapping data from Table 5.2.

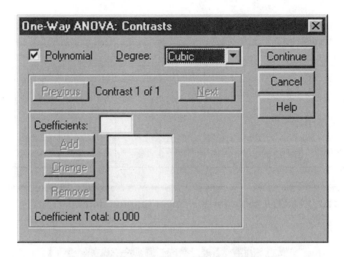

Display 5.9 Generating orthogonal polynomial contrasts in the one-way design.

note that the coefficients defining each trend type sum to zero and so define what is generally referred to as a *contrast* among the factor level means.

SPSS provides trend tests for the one-way design when **Polynomial** is checked on the **One-Way ANOVA: Contrasts** sub-dialogue box (Display 5.9). The types of trend to be evaluated are determined by the **Degree** setting of the box. For our four dose levels, we opted for the largest degree of trend possible, i.e., we asked for trends up to third order (setting **Cubic**). (Note that the more general **Univariate** ANOVA procedure to be discussed in the next Chapter offers more specific factor level comparisons. In addition to **Polynomial Contrasts**, for example, **Simple Contrasts** that compare each group with a reference group, or **Helmert Contrasts** that compare each level with the mean of subsequent factor levels, can be obtained.)

Display 5.10 shows the generated ANOVA table. The rows labeled "Between Groups," "Within Groups," and "Total" simply make up the one-way ANOVA table as discussed in the previous example. More interesting here is the partition of the between-group sum of squares into three single-degree-of-freedom terms corresponding to a linear, a quadratic, and a cubic trend in mean finger tapping frequency over dose. We see that almost all of the between-group sum of squares (186.25 out of 188.08) is accounted for by a linear trend over dose and this trend is statistically significant ($F_{(1,36)} = 41.1, p < 0.001$). The unaccounted part of the between-group sum of squares (1.83) is listed as a deviation from linear trend and allows the construction of a test for trend other than linear. Here we conclude that there is no evidence for a departure from linear trend over dose ($F_{(2,36)} = 0.2, p = 0.82$). So a linear trend suffices to describe the relationship between finger tapping frequency and caffeine dose.

ANOVA

finger taps per minute

			Sum of Squares	df	Mean Square	F	Sig.
Between Groups	(Combined)		188.075	3	62.692	13.821	.000
	Linear Term	Contrast	186.245	1	186.245	41.058	.000
		Deviation	1.830	2	.915	.202	.818
	Quadratic Term	Contrast	.025	1	.025	.006	.941
		Deviation	1.805	1	1.805	.398	.532
	Cubic Term	Contrast	1.805	1	1.805	.398	.532
Within Groups			163.300	36	4.536		
Total			351.375	39			

Display 5.10 ANOVA table for the finger tapping data.

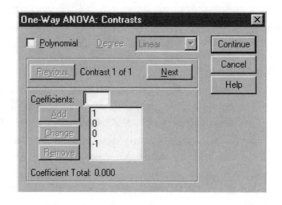

Display 5.11 Generating a user-defined contrast in the one-way design.

The **Contrasts...** sub-dialogue box also lets users define their own contrasts. If, for example, we had simply been interested in a comparison between the no caffeine (factor level 1) and the highest caffeine (factor level 4) groups, we would be testing the null hypothesis

$$1 \times \mu_1 + 0 \times \mu_2 + 0 \times \mu_3 + (-1) \times \mu_4 = 0 \qquad (5.1)$$

where μ_i is the mean finger tapping frequency for factor level i. The corresponding contrast is $\mathbf{c} = (1, 0, 0, -1)$ and can be conveyed to SPSS as shown in Display 5.11.

Display 5.12 gives the resulting output. The "Contrast Coefficients" table reiterates the contrast coefficients and the "Contrasts Tests" table provides an estimate of the contrast, the standard error of that estimator, and a t-test for the null hypothesis that the contrast is zero. Our chosen

Contrast Coefficients

	DOSE			
Contrast	0ml	100ml	200ml	300ml
1	1	0	0	-1

Contrast Tests

		Contrast	Value of Contrast	Std. Error	t	df	Sig. (2-tailed)
finger taps per minute	Assume equal variances	1	-5.60	.952	-5.879	36	.000
	Does not assume equal	1	-5.60	.919	-6.094	15.956	.000

Display 5.12   Contrast output generated by commands in Display 5.11.

contrast represents the difference in mean finger tapping between the no caffeine (0 ml) and the high caffeine consumption (300 ml) groups. This difference is statistically significant ($t(36) = 5.9$, $p < 0.001$) and we estimate that high caffeine consumption increases tapping by 5.6 finger taps per minute relative to no caffeine consumption (standard error = 0.95).

5.3.3 Social Skills of Females

The female social skills data in Table 5.3 is already in the format of a **Data View** spreadsheet. Like the previous two data sets, this data set arises from a one-way design; a single group factor is of interest. But in contrast to the previous examples, four (dependent) social skills variables have been measured on each woman taking part in the study. If the separate variables were of interest, we could apply a *univariate* one-way ANOVA of each, as described previously. Here, however, it is group differences on the underlying concept "social confidence" that are of concern, and therefore we need to carry out a *multivariate* one-way ANOVA.

A one-way design with several dependent variables can be set up in SPSS by using the commands

Analyze – General Linear Model – Multivariate...

The resulting dialogue box is filled in as shown in Display 5.13. This produces the output shown in Display 5.14.

The "Between-Subjects Factors" table in Display 5.14 provides a short design summary. Here we are only investigating a single factor "group," with 11 females observed for each factor level. The "GROUP" part of the "Multivariate Tests" table contains tests for testing the null hypothesis that the concept underlying the multiple dependent variables is not affected by group membership. SPSS automatically lists four commonly used multivariate tests: *Pillai's trace test, Wilks' lambda test, Hotelling's trace test,* and

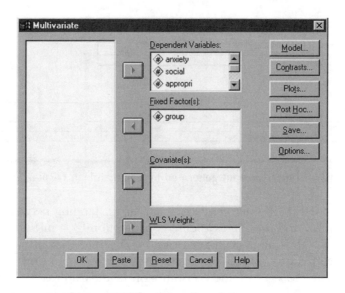

Display 5.13   Defining a one-way design for multiple dependent variables.

Between-Subjects Factors

		Value Label	N
GROUP	1	Behavioural rehearsal	11
	2	Control	11
	3	Behavioural rehearsal + Cognitive restructuring	11

Multivariate Tests[c]

Effect		Value	F	Hypothesis df	Error df	Sig.
Intercept	Pillai's Trace	.998	2838.632[a]	4.000	27.000	.000
	Wilks' Lambda	.002	2838.632[a]	4.000	27.000	.000
	Hotelling's Trace	420.538	2838.632[a]	4.000	27.000	.000
	Roy's Largest Root	420.538	2838.632[a]	4.000	27.000	.000
GROUP	Pillai's Trace	.680	3.604	8.000	56.000	.002
	Wilks' Lambda	.369	4.361[a]	8.000	54.000	.000
	Hotelling's Trace	1.577	5.126	8.000	52.000	.000
	Roy's Largest Root	1.488	10.418[b]	4.000	28.000	.000

[a.] Exact statistic

[b.] The statistic is an upper bound on F that yields a lower bound on the significance level.

[c.] Design: Intercept+GROUP

Display 5.14   MANOVA output for social skills data.

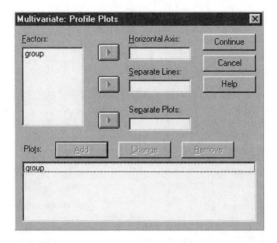

Display 5.15   Generating means plots for multiple dependent variables.

Roy's largest root test (details of all these multivariate tests can be found in Stevens, 1992). The different test statistic may give different results when used on the same set of data, although the resulting conclusion from each is often the same. In practice, one would decide in advance for a particular multivariate test. For each multivariate procedure, the initial test statistic (given under "Value") is transformed into a test statistic (given under "F"), which can be compared with an *F*-distribution with "Hypothesis df" and "Error df" to derive the p-value of the test (given under "Sig."). In our data example, all four multivariate tests lead to the same conclusion; the set of four social skills variables *is* affected by group membership (Wilks' lambda test: $F(8, 54) = 4.4$, $p < 0.001$).

(Also automatically provided is a table labeled "Tests of Between-Subjects Effects." This is a compilation of univariate one-way ANOVAs, one for each dependent variable and should be ignored when one has opted for simultaneous testing of the set of dependent variables.)

In order to interpret these results, it will be helpful to look at the mean value of each of the four variables for each group by employing the **Profile Plots** sub-dialogue box to create a means plot (Display 5.15). The resulting output shows that the overall group difference is due to the control group deviating from the two experimental groups, with the experimental groups showing improvements on all four scales (Display 5.16).

The buttons on the **Multivariate** dialogue box shown in Display 5.13 offer similar options to those of the **Univariate** dialogue box, the most important of which will be discussed in the next chapter. However, it is important to note that all pairwise group comparisons offered by the

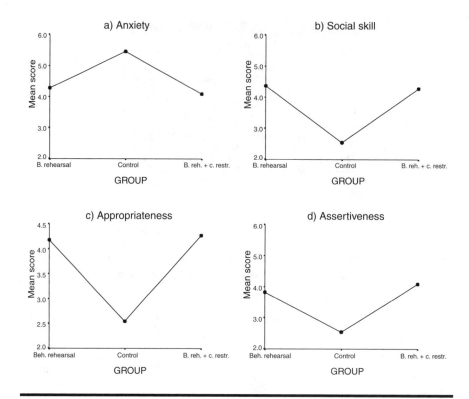

Display 5.16   Means plots for each of the social skills variables.

MANOVA routine (planned or otherwise) are based on separate univariate one-way ANOVAs of the dependent variables and should not be used when one has opted for simultaneous testing.

(SPSS provides a formal test of the common covariance matrix assumption of multivariate analysis of variance, but the test is of little practical value and we suggest that it is not used in practice. Informal assessment of the assumption is preferred, see Exercise 5.4.3.)

5.4 Exercises

5.4.1 *Cortisol Levels in Psychotics: Kruskal-Wallis Test*

Rothschild et al. (1982) report a study of the dexamethasone suppression test as a possible discriminator among subtypes of psychotic patients. Postdexamethasone cortisol levels (in micrograms/dl) in 31 control subjects and four groups of psychotics were collected and are given in Table 5.4.

Table 5.4   Postdexamethasone Cortisol Levels

Control	Major Depression	Bipolar Depression	Schizophrenia	Atypical
1.0	1.0	1.0	0.5	0.9
1.0	3.0	1.0	0.5	1.3
1.0	3.5	1.0	1.0	1.5
1.5	4.0	1.5	1.0	4.8
1.5	10.0	2.0	1.0	
1.5	12.5	2.5	1.0	
1.5	14.0	2.5	1.5	
1.5	15.0	5.5	1.5	
1.5	17.5		1.5	
1.5	18.0		2.0	
1.5	20.0		2.5	
1.5	21.0		2.5	
1.5	24.5		2.5	
2.0	25.0		11.2	
2.0				
2.0				
2.0				
2.0				
2.0				
2.0				
2.0				
2.5				
2.5				
3.0				
3.0				
3.0				
3.5				
3.5				
4.0				
4.5				
10.0				

Source: Hand et al., 1994.

Use one-way analysis of variance to assess whether there is any difference between the five groups in average cortisol level.

A nonparametric alternative to one-way ANOVA is given by the *Kruskal-Wallis test*. This extends the Mann-Whitney U-test described in Chapter 2, Box 2.2, to more than two groups. Use some initial graphical analyses of the data to assess whether the assumptions of the one-way ANOVA are justified for these data. If in doubt, carry out a nonparametric test.

(Hint: In SPSS, the Kruskal-Wallis test can be generated using the commands Analyze – Nonparametric Tests – K Independent Samples...)

5.4.2 Cycling and Knee-Joint Angles

The data given in Table 5.5 were collected in an investigation described by Kapor (1981) in which the effect of knee-joint angle on the efficiency of cycling was studied. Efficiency was measured in terms of distance pedaled on an ergocycle until exhaustion. The experimenter selected three knee-joint angles of particular interest: 50, 70, and 90. Thirty subjects were available for the experiment and ten subjects were randomly allocated to each angle. The drag of the ergocycle was kept constant and subjects were instructed to pedal at a constant speed of 20 km/h.

1. Carry out an initial graphical inspection of the data to assess whether there are any aspects of the observations that might be a cause for concern in later analyses.
2. Derive the appropriate analysis of variance table for the data.
3. Investigate the mean differences between knee-joint angles in more detail using suitable multiple comparison tests.

Table 5.5 The Effect of Knee-Joint Angle on the Efficiency of Cycling: Total Distance Covered (km)

Knee-Joint Angle		
50	70	90
8.4	10.6	3.2
7.0	7.5	4.2
3.0	5.1	3.1
8.0	5.6	6.9
7.8	10.2	7.2
3.3	11.0	3.5
4.3	6.8	3.1
3.6	9.4	4.5
8.0	10.4	3.8
6.8	8.8	3.6

5.4.3 More on Female Social Skills: Informal Assessment of MANOVA Assumptions

The assumptions underlying the MANOVA of the female social skills data from Table 5.3 were not assessed in the main body of the text. Use informal methods to assess:

1. Approximate normality (note that the data cannot, strictly speaking, arise from a normal distribution, since we can only observe integer scores)
2. Homogeneity of covariance matrices

Hints:

1. Use commands Analyze – General Linear Model – Multivariate... – Save... to generate studentized residuals for each dependent variable and assess their distributional shape by a set of appropriate diagnostic plots.
2. Use commands Data – Split File... to Organize output by groups with Groups Based on: group so that all further commands are carried out within subject groups. Then, for each group, generate standard deviations for each dependent variable and a correlation matrix, and compare values across the three groups.

Chapter 6

Analysis of Variance II: Factorial Designs; Does Marijuana Slow You Down? and Do Slimming Clinics Work?

6.1 Description of Data

Two data sets will be of concern in this chapter. The first, shown in Table 6.1, arises from a study investigating whether the effects of marijuana vary with prior usage of the drug. To address this question, an experiment involving 12 moderate users of the drug, 12 light users, and 12 students who had had no prior use of marijuana was carried out. All participants were randomly sampled from a college population. Within each usage level, half of the students were randomly assigned to a placebo condition and the other half to an experimental condition. In the placebo condition, each subject smoked two regular cigarettes that tasted and smelled like marijuana cigarettes. In the experimental condition, each subject smoked two marijuana cigarettes. Immediately after smoking, each subject was given a reaction time test.

Table 6.1   Reaction Times (Milliseconds) from Study Investigating the Effects of Marijuana

Previous Marijuana Usage	Placebo		Experimental	
None	795	605	965	878
	700	752	865	916
	648	710	811	840
Light	800	610	843	665
	705	757	765	810
	645	712	713	776
Moderate	790	600	815	635
	695	752	735	782
	634	705	983	744

Source: Pagano, 1998, with permission of the publisher.

The second data set for this chapter is shown in Table 6.2. These data arise from an investigation into the effectiveness of slimming clinics. Such clinics aim to help people lose weight by offering encouragement and support about dieting through regular meetings. Two factors were examined in the study:

- *Technical manual*: This contained slimming advice based on psychological behavior theory. Some slimmers were given the manual and others were not.
- *Previous slimmer*: Had the slimmer already been trying to slim or not?

The response variable in the study was

$$\text{response} = \frac{\text{weight at three months} - \text{ideal weight}}{\text{initial weight} - \text{ideal weight}} \qquad (6.1)$$

Both data sets come from studies employing a *two-way design*; they differ in that in Table 6.1 there are an equal number of observations in each cell of the table, whereas in Table 6.2, this is not the case. This difference has implications for the analysis of each data set as we shall see later in the chapter.

6.2 Analysis of Variance

To investigate each of the two data sets described in the previous section, we shall use a *two-way analysis of variance*. The model on which such

Table 6.2 Data from Slimming Clinics

Manual	Experience	Response	Manual	Experience	Response
1	1	−14.67	1	1	−1.85
1	1	−8.55	1	1	−23.03
1	1	11.61	1	2	0.81
1	2	2.38	1	2	2.74
1	2	3.36	1	2	2.10
1	2	−0.83	1	2	−3.05
1	2	−5.98	1	2	−3.64
1	2	−7.38	1	2	−3.60
1	2	−0.94	2	1	−3.39
2	1	−4.00	2	1	−2.31
2	1	−3.60	2	1	−7.69
2	1	−13.92	2	1	−7.64
2	1	−7.59	2	1	−1.62
2	1	−12.21	2	1	−8.85
2	2	5.84	2	2	1.71
2	2	−4.10	2	2	−5.19
2	2	0.00	2	2	−2.80

Source: Hand et al., 1994.

an analysis is based and some other aspects of analysis of variance are described briefly in Box 6.1. (Full details are given in Everitt, 2001b.) Essentially, application of analysis of variance to these two data sets will enable us to answer questions about mean differences between populations defined by either factor and, in addition, to assess whether there is any *interaction* between the two factors, i.e., whether or not the separate effects of each factor on the response are additive.

Box 6.1 Two-Way Analysis of Variance

■ A suitable model for a general two-way design is the following:

$$y_{ijk} = \mu + \alpha_i + \beta_j + \alpha\beta_{ij} + \varepsilon_{ijk}$$

where y_{ijk} represents the value of the response variable for the *k*th subject in the *i*th level of say factor A (with, say, *a* levels) and the *j*th level of factor B (with, say, *b* levels).

■ In this model, μ represents the overall mean, α_i the effect of level i of factor A, β_j the effect of level j of factor B, and $\alpha\beta_{ij}$ the interaction effect. The residual or error terms, ε_{ijk}, are assumed normally distributed with mean zero, and variance σ^2.

■ The model as written is *overparameterized*, so the following constraints are generally applied to the parameters to overcome the problem (see Maxwell and Delaney, 1990, for details):

$$\sum_i \alpha_i = 0; \quad \sum_j \beta_j = 0; \quad \sum_i \alpha\beta_{ij} = \sum_j \alpha\beta_{ij} = 0$$

■ The null hypotheses of interest can be written in terms of the parameters in the model as

$$H_0^{(1)}: \alpha_1 = \alpha_2 = \ldots = \alpha_a = 0 \quad \text{(no main effect of factor A)}$$

$$H_0^{(2)}: \beta_1 = \beta_2 = \ldots = \beta_b = 0 \quad \text{(no main effect of factor B)}$$

$$H_0^{(3)}: \alpha\beta_{11} = \alpha\beta_{12} = \ldots = \alpha\beta_{ab} = 0 \quad \text{(no interaction effect)}$$

■ The total variation in the observations is partitioned into that due to differences between the means of the levels of factor A, that due to differences between the means of the levels of factor B, that due to the interaction of A and B, and that due to differences among observations in the same factor level combination (cell). A *two-way analysis of variance table* can be constructed as follows:

Source	DF	SS	MS	MSR (F)
Between cells				
Between factor A levels	$a-1$	SSA	$SSA/(a-1)$	MSA/MSE
Between factor B levels	$b-1$	SSB	$SSB/(b-1)$	MSB/MSE
Factor A × factor B	$(a-1)(b-1)$	SSAB	$SSAB/[(a-1)(b-1)]$	MSAB/MSE
Within cells	$N-ab$	SSE	$SSE/(N-ab)$	
Total	$N-1$			

■ Under each of the null hypotheses above, the respective mean square ratio (MSR) has an F distribution with model term and error degrees of freedom.

- When the cells of the design have different numbers of observation, it is not possible to partition the between cell variation into nonoverlapping parts corresponding to factor main effects and an interaction. We shall say more about this when analyzing the slimming clinic data.
- The *F*-tests are valid under the following assumptions:
  1. The observations in each cell come from a normal distribution.
  2. The population variances of each cell are the same.
  3. The observations are independent of one another.

6.3 Analysis Using SPSS

6.3.1 Effects of Marijuana Use

The data in Table 6.1 arise from a two-way design, with factors "past marijuana use" (three levels: "no use," "light use," "moderate use") and "current marijuana use" (two levels: "placebo group," "experimental group") and six replicate measures of reaction time per factor level combination. To convey the factorial structure of the experiment to SPSS, a **Data View** spreadsheet needs to be set up as shown in Display 6.1. The rows of the spreadsheet correspond to subjects, and for each subject we have recorded the level of past (**past_use**) and current marijuana use (**group**) and the observed reaction time (**time**).

Informative summary statistics and the required analysis of variance table for the data can be obtained by using the general ANOVA routine in SPSS

Analyze – General Linear Model – Univariate...

The **Univariate** dialogue box is filled in as shown in Display 6.2 to define the correct structure for the marijuana data. (Both factors in this experiment are regarded as *fixed;* for a discussion of both fixed and random factors, see Everitt, 2001b.) Within-group summary, statistics can then be requested by checking **Descriptive statistics** in the **Options** sub-dialogue box.

The resulting descriptive output is shown in Display 6.3. For each of the six factor level combinations, SPSS displays the mean, standard deviation, and number of replicates. It is apparent from these figures that within each category of previous marijuana usage, the students who just smoked two marijuana cigarettes have slower reaction times than those who used the regular cigarettes. Across previous marijuana use levels, however, the situation is not as clear.

	subject	past_use	group	time
13	13	2	1	610
14	14	2	1	757
15	15	2	1	712
16	16	3	1	600
17	17	3	1	752
18	18	3	1	705
19	19	1	2	965
20	20	1	2	865
21	21	1	2	811
22	22	2	2	843
23	23	2	2	765
24	24	2	2	713
25	25	3	2	815
26	26	3	2	735
27	27	3	2	983
28	28	1	2	878
29	29	1	2	916
30	30	1	2	840
31	31	2	2	665
32	32	2	2	810
33	33	2	2	776
34	34	3	2	635
35	35	3	2	782
36	36	3	2	744
37				

Display 6.1   SPSS spreadsheet containing data from Table 6.1.

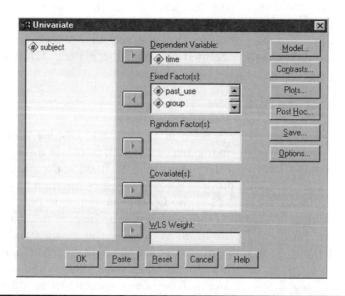

Display 6.2   Defining a two-way design.

Descriptive Statistics

Dependent Variable: reaction time

past marijuana use	current marijuana use	Mean	Std. Deviation	N
none	Smoked two placebo cigarettes	701.67	68.617	6
	Smoked two Marijuana cigarettes	879.17	54.967	6
	Total	790.42	110.028	12
light	Smoked two placebo cigarettes	704.83	69.861	6
	Smoked two Marijuana cigarettes	762.00	64.622	6
	Total	733.42	70.767	12
moderate	Smoked two placebo cigarettes	696.00	70.872	6
	Smoked two Marijuana cigarettes	782.33	115.543	6
	Total	739.17	101.903	12
Total	Smoked two placebo cigarettes	700.83	65.663	18
	Smoked two Marijuana cigarettes	807.83	93.863	18
	Total	754.33	96.527	36

Display 6.3   Descriptive statistics for reaction times.

A design summary table and the analysis of variance table resulting from the previous commands are shown in Display 6.4. The rows of the "Tests of Between-Subjects Effects" table labeled "PAST_USE," "GROUP," "PAST_USE*GROUP," "Error," and "Corrected Total" make up the two-way ANOVA table as described in Box 6.1. (The remaining rows of the table are not generally of interest in a factorial design.) SPSS output commonly employs the "*" notation to denote an interaction effect between two variables. We find no statistical evidence of an interaction between current and previous marijuana use on reaction times ($F(2,30) = 2$, $p = 0.15$); thus, the effect of smoking marijuana on reaction time does not vary significantly with prior usage. Also there is no significant main effect of previous marijuana use ($F(2,30) = 2$, $p = 0.15$), so that the mean reaction times of the previous marijuana use groups do not differ significantly. There is, however, a significant main effect of factor **group** ($F(1,30) = 17.6$, $p < 0.001$), implying that mean reaction times differ between the placebo and the experimental group. Finally, the table provides R^2 and its adjusted version as a measure of model fit. Our fitted ANOVA model is able to explain 46.1% of the variance in reaction times.

This is a convenient point at which to introduce a graphical display that is often useful in understanding interactions (or the lack of them) in two-way and other factorial designs. The display is essentially a plot of estimated cell means. SPSS refers to such plots as "profile plots" and provides them via the **Plots...** button on the **Univariate** dialogue box (see

Between-Subjects Factors

		Value Label	N
past marijuana use	1	none	12
	2	light	12
	3	moderate	12
current marijuana use	1	Smoked two placebo cigarettes	18
	2	Smoked two Marijuana cigarettes	18

Tests of Between-Subjects Effects

Dependent Variable: reaction time

Source	Type III Sum of Squares	df	Mean Square	F	Sig.
Corrected Model	150317.667[a]	5	30063.533	5.130	.002
Intercept	20484676.0	1	20484676.00	3495.751	.000
PAST_USE	23634.500	2	11817.250	2.017	.151
GROUP	103041.000	1	103041.000	17.584	.000
PAST_USE * GROUP	23642.167	2	11821.083	2.017	.151
Error	175796.333	30	5859.878		
Total	20810790.0	36			
Corrected Total	326114.000	35			

a. R Squared = .461 (Adjusted R Squared = .371)

Display 6.4 ANOVA table for reaction times.

Display 6.2). The resulting **Profile Plots** sub-dialogue box is shown in Display 6.5. The user defines the factor levels to be displayed on the *x*-axis or as separate lines and then presses **Add** to store the graph instruction. Several different displays can be requested. Here we opt for displaying the previous marijuana use categories on the *x*-axis and the current use groups as separate lines (Display 6.5). The resulting graph is shown in Display 6.6. An interaction between the two factors would be suggested by nonparallel lines in this diagram. Here we can see that there is some departure from parallel lines; the reaction times of those students who never used marijuana in the past appear to increase more under the effect of smoking marijuana than those who were previous users. However, as the ANOVA table shows, this degree of observed interaction is consistent with the absence on an interaction in the population ($p = 0.15$).

Rather than relying on p-values to identify important effects, it is generally advisable to quantify these effects by constructing appropriate confidence intervals. Here, with a significant main effect of **group** and a nonsignificant interaction effect, it is sensible to estimate the difference

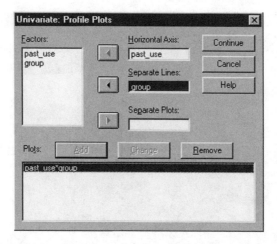

Display 6.5   Requesting a line chart of estimated mean responses.

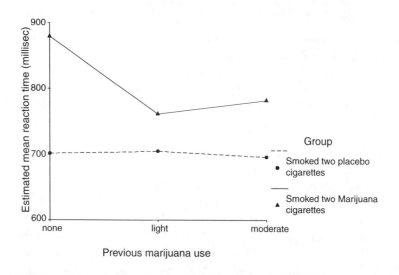

Display 6.6   Line chart of estimated means for reaction times.

in mean reaction times between the experimental and control group. We can request this confidence interval by using the **Options** sub-dialogue box as shown in Display 6.7. This generates two further output tables shown in Display 6.8.

The "Estimates" table provides estimates, standard errors of estimators, and 95% confidence intervals for the mean reaction times in the placebo and the experimental group. Here the estimated means are the same as

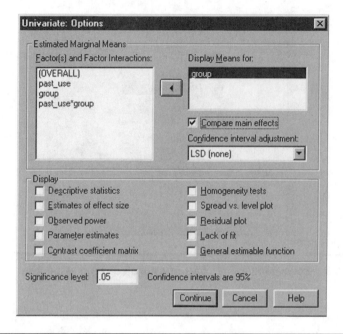

Display 6.7   Requesting comparisons between the levels of a factor.

those shown in the "Descriptive Statistics" table (Display 6.3), but this is not necessarily always the case, as we will see in the analysis of the second example. (Note also that the confidence intervals given in Display 6.8 are based on estimating the assumed common within-group variance.)

The "Pairwise Comparisons" table provides estimates of differences between group means, their standard errors, a test of zero difference, and, finally, a confidence intervals for the difference. SPSS repeats each group comparison twice depending on the direction of the difference. Here we simply consider the first row of the table. The difference in mean reaction times between the placebo group (I) and the experimental group (J) is estimated as −107 milliseconds; after smoking two marijuana cigarettes subjects are estimated to be 107 milliseconds slower on a reaction time task irrespective of previous marijuana use (95% CI from 55 to 159 milliseconds).

Like the **One-way ANOVA** dialogue box, the **Univariate** dialogue box contains buttons labeled **Post Hoc...** and **Contrasts...** for requesting *post hoc* and planned contrast comparisons, respectively (see Display 6.2). However, it is important to note that SPSS only offers comparisons (planned or otherwise) between the levels of a single factor. In a factorial design, the comparison procedures should, therefore, only be used when the factors do *not* interact with each other so that interpreting main effects of factors is appropriate.

Estimates

Dependent Variable: reaction time

current marijuana use	Mean	Std. Error	95% Confidence Interval	
			Lower Bound	Upper Bound
Smoked two placebo cigarettes	700.833	18.043	663.985	737.682
Smoked two Marijuana cigarettes	807.833	18.043	770.985	844.682

Pairwise Comparisons

Dependent Variable: reaction time

(I) current marijuana use	(J) current marijuana use	Mean Difference (I-J)	Std. Error	Sig.[a]	95% Confidence Interval for Difference[a]	
					Lower. Bound	Upper Bound
Smoked two placebo cigarettes	Smoked two Marijuana cigarettes	-107.000*	25.517	.000	-159.112	-54.888
Smoked two Marijuana cigarettes	Smoked two placebo cigarettes	107.000*	25.517	.000	54.888	159.112

Based on estimated marginal means

*. The mean difference is significant at the .05 level.

a. Adjustment for multiple comparisons: Least Significant Difference (equivalent to no adjustments).

Display 6.8 Confidence intervals for group differences in reaction times.

Finally, as in linear regression, it is important to check the assumptions of the underlying linear model. Normality of the error term is best assessed by saving the studentized residual using the **Save** sub-dialogue box and then employing a suitable diagnostic on the generated residual variable (see Chapter 4). A box plot of the studentized residual is shown in Display 6.9. There appears to be no departure from normality, although subject 27 (the subject with the longest reaction time) is labeled an "outlier" by SPSS.

Homogeneity of variance can be assessed by any measure of variability within the factor level combinations. We employ the within-group standard deviations shown in Display 6.3. Except, perhaps, for the group of students who previously made moderate use of marijuana and smoked two marijuana cigarettes before the task, the variances appear fairly similar.

6.3.2 Slimming Clinics

The slimming data in Table 6.2 are already in the appropriate format for applying analysis of variance in SPSS. The corresponding **Data View** spreadsheet contains participants' weight change indices (column labeled **response**) together with categories indicating the use of a manual (**manual**) and previous experience (**experien**). As in the previous example, interest centers on investigating main effects of the factors "manual use" and

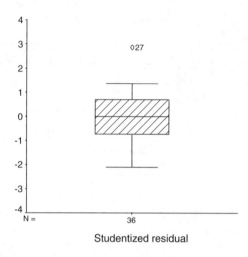

Display 6.9   Box plot of studentized residuals from two-way ANOVA of reaction times.

"slimming experience" on weight loss and assessing whether there is any evidence of an interaction between the two factors.

However, in contrast to the previous data set, this two-way design is not balanced. Rather, the number of subjects varies between the factor level combinations. We can see this easily by constructing a cross-tabulation of the factors (Display 6.10). Clearly, in our sample an association existed between the two factors with the majority of the subjects with previous slimming experience (69%) not using a manual and the majority of the slimming novices (67%) using a manual.

Similar to the lack of independence of explanatory variables in the regression setting (see Chapter 4), lack of balance between the factors of a multi-way ANOVA has the effect that the amount of variability that is explained by a model term depends on the terms already included in the model. For the slimming data, this means that the amount of variability in the weight index that can be explained by the factor **experien** depends on whether or not the factor **manual** is already included in the model and vice versa.

We can see this by asking SPSS to fit the model terms in different orders. SPSS recognizes the order of the factors when the **Sum of squares** option on the in the **Univariate: Model** sub-dialogue box is set to **Type I** (Display 6.11). The factor order chosen in Display 6.11 produces the ANOVA table shown in Display 6.12. We then switched the order of the factors in the **Fixed factor(s)** list of the **Univariate** dialogue box so that **experien** was considered before **manual**. This produced the ANOVA table shown in Display 6.13. We find that when factor **manual** is fitted first, it accounts for

MANUAL * EXPERIEN Crosstabulation

			EXPERIEN		
			experienced	novice	Total
MANUAL	manual	Count	5	12	17
		% within EXPERIEN	31.3%	66.7%	50.0%
	no manual	Count	11	6	17
		% within EXPERIEN	68.8%	33.3%	50.0%
Total		Count	16	18	34
		% within EXPERIEN	100.0%	100.0%	100.0%

Display 6.10   Cross-tabulation of manual use by slimming experience.

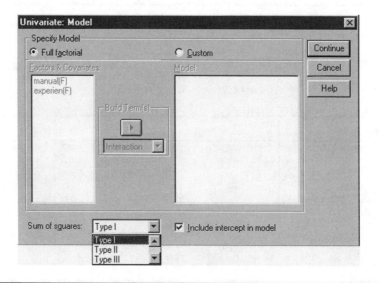

Display 6.11   Setting the type of sum of squares.

a sum of squares of 21.2 (Display 6.12). By contrast, when **manual** is fitted after **experience**, so that only the sum of squares that cannot be attributed to **experien** is attributed to **manual**, then the sum of squares is reduced to 2.1 (Display 6.13). A similar effect is observed for factor **experien**. When it is fitted first, it is attributed a larger sum of squares (285, Display 6.13) than when it is fitted after **manual** is already included in the model (265.9, Display 6.12).

So, how should we analyze unbalanced factorial designs? There has been considerable discussion on this topic centering on how the sum of squares should be attributed to a factor. There are three basic approaches to attributing the sum of squares (SS), called types I, II, and III in SPSS (there is also a type IV which need not concern us here):

Tests of Between-Subjects Effects

Dependent Variable: RESPONSE

Source	Type I Sum of Squares	df	Mean Square	F	Sig.
Corrected Model	287.232[a]	3	95.744	2.662	.066
Intercept	480.979	1	480.979	13.375	.001
MANUAL	21.188	1	21.188	.589	.449
EXPERIEN	265.914	1	265.914	7.394	.011
MANUAL * EXPERIEN	.130	1	.130	.004	.952
Error	1078.848	30	35.962		
Total	1847.059	34			
Corrected Total	1366.080	33			

a. R Squared = .210 (Adjusted R Squared = .131)

Display 6.12 Two-way ANOVA using type I sum of squares, factor manual fitted before factor experien.

Tests of Between-Subjects Effects

Dependent Variable: RESPONSE

Source	Type I Sum of Squares	df	Mean Square	F	Sig.
Corrected Model	287.232[a]	3	95.744	2.662	.066
Intercept	480.979	1	480.979	13.375	.001
EXPERIEN	284.971	1	284.971	7.924	.009
MANUAL	2.130	1	2.130	.059	.809
EXPERIEN * MANUAL	.130	1	.130	.004	.952
Error	1078.848	30	35.962		
Total	1847.059	34			
Corrected Total	1366.080	33			

a. R Squared = .210 (Adjusted R Squared = .131)

Display 6.13 Two-way ANOVA using type I sum of squares, factor experien fitted before factor manual.

■ *Type I SS* (also known as "Hierarchical method") fits the factors in the order that they are supplied. Interaction terms are always considered last. As a result, the SS of the model terms and the error SS add up to the (corrected) total SS.

■ *Type II SS* (also known as "Experimental method") is similar to type III except that it adjusts the main effects only for other main effects and not for interaction terms.

■ *Type III SS* (also known as "Unique method") adjusts each model term for all the other model terms. The term SS and the error SS do not add up.

In a balanced design, the type I, II, or III SS for corresponding terms are identical and the issue of choosing the type of SS does not arise. For the unbalanced design, authors such as Maxwell and Delaney (1990) and Howell (2002) recommend the use of type III SS, and these are the default setting in SPSS. Nelder (1977) and Aitkin (1978), however, are strongly critical of "correcting" main effect SS for an interaction term involving the corresponding main effects. The arguments are relatively subtle, but for a two-factor design (factors A and B) they go something like this:

1. When models are fitted to data, the principle of *parsimony* is of critical importance. In choosing among possible models, we do not adopt complex models for which there is no empirical evidence.
2. So, if there is no convincing evidence of an AB interaction, we do not retain the term in the model. Thus, additivity of A and B is assumed unless there is convincing evidence to the contrary.
3. The issue does not arise as clearly in the balanced case, for there the sum of squares for A, for example, is independent of whether interaction is assumed or not. Thus, in deciding on possible models for the data, we do not include the interaction term unless it has been shown to be necessary, in which case tests on main effects involved in the interaction are not carried out (or if carried out, not interpreted).
4. Thus, the argument proceeds that type III sum of squares for A, in which it is adjusted for AB as well as B, makes no sense.
5. First, if the interaction term is necessary in the model, then the experimenter will usually wish to consider simple effects of A at each level of B separately. A test of the hypothesis of no A main effect would not usually be carried out if the AB interaction is significant.
6. If the AB interaction is not significant, then adjusting for it is of no interest and causes a substantial loss of power in testing the A and B main effects.

Display 6.14 shows the ANOVA table generated by choosing type II sum of squares. Basically, the table is a compilation of the ANOVA tables in Displays 6.12 and 6.13. The row for factor **experien** is taken from the table in Display 6.12 and that for factor **manual** from the table in Display 6.13. We finally conclude that there is no statistical evidence for an interaction effect of experience and manual use on weight loss ($F(1,30) = 0.004$, $p = 0.95$). The response variable is also not significantly affected by manual use ($F(1,30) = 0.06$, $p = 0.81$), but there is evidence of an effect of previous slimming experience ($F(1,30) = 7.4$, $p = 0.011$).

Tests of Between-Subjects Effects

Dependent Variable: RESPONSE

Source	Type II Sum of Squares	df	Mean Square	F	Sig.
Corrected Model	287.232[a]	3	95.744	2.662	.066
Intercept	480.979	1	480.979	13.375	.001
EXPERIEN	265.914	1	265.914	7.394	.011
MANUAL	2.130	1	2.130	.059	.809
EXPERIEN * MANUAL	.130	1	.130	.004	.952
Error	1078.848	30	35.962		
Total	1847.059	34			
Corrected Total	1366.080	33			

a. R Squared = .210 (Adjusted R Squared = .131)

Display 6.14   Two-way ANOVA of weight losses using type II sum of squares.

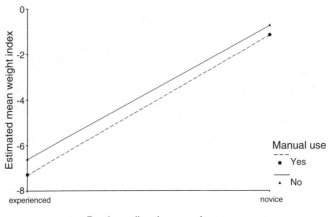

Display 6.15   Line chart of estimated mean weight loss indices.

The line chart of estimated mean responses generated using the Profile Plots sub-dialogue box (see Display 6.5) is shown in Display 6.15 and immediately explains the finding. In our sample, using a manual reduced the weight only minimally and the reduction did not vary with previous experience. Previous experience, on the other hand, reduced the index considerably. A reduction estimate of 6, with a 95% CI from 1.5 to 10.5, was obtained using the Options sub-dialogue box (see Display 6.7). Note that in the unbalanced case, this mean difference estimate is no longer simply the difference in observed group means (which would be 5.8).

(We leave it as an exercise for the reader, see exercise 6.4.4, to check whether or not the assumptions of the ANOVA model for these data are reasonable.)

6.4 Exercises

6.4.1 Headache Treatments

The data shown below come from an investigation into the effectiveness of different kinds of psychological treatment on the sensitivity of headache sufferers to noise. There are two groups of 22 subjects each: those suffering from a migraine headache (coded 1 in the table) and those suffering from a tension headache (coded 2). Half of each of the two headache groups were randomly selected to receive a placebo treatment (coded 1), and the other half an active treatment (coded 2). Each subject rated his or her own noise sensitivity on a visual analogue scale ranging from low to high sensitivity. Investigate whether the treatment is effective and whether there is any evidence of a treatment × type of headache interaction.

Table 6.3   Psychological Treatments for Headaches

Headache Type	Treatment Group	Score	Headache Type	Treatment Group	Score
1	1	5.70	2	1	2.70
1	1	5.63	2	1	4.65
1	1	4.83	2	1	5.25
1	1	3.40	2	1	8.78
1	1	15.20	2	1	3.13
1	1	1.40	2	1	3.27
1	1	4.03	2	1	7.54
1	1	6.94	2	1	5.12
1	1	0.88	2	1	2.31
1	1	2.00	2	1	1.36
1	1	1.56	2	1	1.11
1	2	2.80	2	2	2.10
1	2	2.20	2	2	1.42
1	2	1.20	2	2	4.98
1	2	1.20	2	2	3.36
1	2	0.43	2	2	2.44
1	2	1.78	2	2	3.20
1	2	11.50	2	2	1.71
1	2	0.64	2	2	1.24
1	2	0.95	2	2	1.24
1	2	0.58	2	2	2.00
1	2	0.83	2	2	4.01

6.4.2 Biofeedback and Hypertension

Maxwell and Delaney (1990) describe a study in which the effects of three possible treatments for hypertension were investigated. The details of the treatments are as follows:

Table 6.4   Three-Way Design

Treatment	Description	Levels
Drug	Medication	Drug X, Drug Y, Drug Z
Biofeed	Psychological feedback	Present, absent
Diet	Special diet	Given, not given

All 12 combinations of the three treatments were included in a $3 \times 2 \times 2$ design. Seventy-two subjects suffering from hypertension were recruited to the study, with six being randomly allocated to each of the 12 treatment combinations. Blood pressure measurements were made on each subject after treatments, leading to the data in Table 6.5. Apply analysis of variance to these data and interpret your results using any graphical displays that you think might be helpful.

Table 6.5   Blood Pressure Data

Treatment							Special Diet					
Biofeedback	Drug			No						Yes		
Present	X	170	175	165	180	160	158	161	173	157	152	181	190
	Y	186	194	201	215	219	209	164	166	159	182	187	174
	Z	180	187	199	170	204	194	162	184	183	156	180	173
Absent	X	173	194	197	190	176	198	164	190	169	164	176	175
	Y	189	194	217	206	199	195	171	173	196	199	180	203
	Z	202	228	190	206	224	204	205	199	170	160	179	179

6.4.3 Cleaning Cars Revisited: Analysis of Covariance

Analysis of covariance (ANCOVA) is typically concerned with comparing a continuous outcome between the levels of a factor while adjusting for the effect of one or more continuous explanatory variables.

Both, ANOVA as well as regression, are special cases of linear models, and hence ANCOVA can be viewed as an extension of ANOVA to include continuous explanatory variables (often called "covariates") or an extension of regression to include categorical variables. Note that standard

ANCOVA assumes that there is no interaction between the continuous and categorical explanatory variables; i.e., the effect of the continuous variables is assumed constant across the groups defined by the categorical explanatory variables. More details on the use of ANCOVA can be found in Everitt (2001b).

The data on car cleaning in Table 4.1 were previously analyzed by a multiple linear regression of the dependent variable, car cleaning time, on the two explanatory variables, extroversion rating and gender (see Chapter 4). We showed that categorical variables, such as gender, could be accommodated within the regression framework by representing them by dummy variables. This regression analysis was, in fact, an ANCOVA of the car cleaning times.

The **Univariate** dialogue box can accommodate continuous explanatory variables, such as extroversion rating, in an ANOVA model by including them in the **Covariate(s)** list. Provided the type of sums of squares is left as its default setting (Type III), every term in the resulting "Tests of Between-Subjects Effects" table is then adjusted for other terms in the model.

Regression coefficients can also be requested by checking **Parameter estimates** in the **Options** sub-dialogue box. This results in a further output table titled "Parameter Estimates" which resembles the "Coefficients" table created by the regression routine. (Note that when categorical variables are specified as fixed factors in the **Univariate** dialogue box, they are automatically dummy-coded so that the category with the highest code represents the reference category. This affects the interpretation of the level difference parameters and of the intercept parameter.)

Use the **Univariate** dialogue box to carry out an ANCOVA of car cleaning times and regenerate the results obtained via regression modeling (cf. Model 2 in the "Model Summary" and "Coefficients" tables in Chapter 4, Display 4.8).

6.4.4 More on Slimming Clinics

Assess the assumptions underlying the two-way ANOVA of the slimming data in Table 6.2 informally using appropriate diagnostic plots.

Chapter 7

Analysis of Repeated Measures I: Analysis of Variance Type Models; Field Dependence and a Reverse Stroop Task

---

7.1 Description of Data

The data of concern in this chapter are shown in Table 7.1. Subjects selected at random from a large group of potential subjects identified as having field-independent or field-dependent cognitive style were required to read two types of words (color and form names) under three cue conditions — normal, congruent, and incongruent. The order in which the six reading tasks were carried out was randomized. The dependent variable was the time in milliseconds taken to read the stimulus words.

7.2 Repeated Measures Analysis of Variance

The observations in Table 7.1 involve *repeated measures* of the response variable on each subject. Here six repeated measurements have been

171

Table 7.1   Field Independence and a Reverse Stroop Task

	Form Names			Color Names		
Subject	Normal Condition	Congruent Condition	Incongruent Condition	Normal Condition	Congruent Condition	Incongruent Condition
Field-independent						
1	191	206	219	176	182	196
2	175	183	186	148	156	161
3	166	165	161	138	146	150
4	206	190	212	174	178	184
5	179	187	171	182	185	210
6	183	175	171	182	185	210
7	174	168	187	167	160	178
8	185	186	185	153	159	169
9	182	189	201	173	177	183
10	191	192	208	168	169	187
11	162	163	168	135	141	145
12	162	162	170	142	147	151
Field-dependent						
13	277	267	322	205	231	255
14	235	216	271	161	183	187
15	150	150	165	140	140	156
16	400	404	379	214	223	216
17	183	165	187	140	146	163
18	162	215	184	144	156	165
19	163	179	172	170	189	192
20	163	159	159	143	150	148
21	237	233	238	207	225	228
22	205	177	217	205	208	230
23	178	190	211	144	155	177
24	164	186	187	139	151	163

Note: Response variable is time in milliseconds.

made, each measurement corresponding to a particular combination of type of word and cue condition. Researchers typically adopt the repeated measures paradigm as a means of reducing error variability and as the natural way of measuring certain phenomena (for example, developmental changes over time, and learning and memory tasks). In this type of design, effects of experimental factors giving rise to the repeated measures are assessed relative to the average response made by the subject on all conditions or occasions. In essence, each subject serves as his or her own control, and, accordingly, variability caused by differences in average responsiveness of the subjects is eliminated from the extraneous error variance. A consequence of this is that the power to detect the effects of *within-subjects* experimental factors (type of word and cue condition in Table 7.1) is increased compared with testing in a between-subjects design.

Unfortunately, the advantages of a repeated measures design come at a cost, namely, the probable lack of independence of the repeated measurements. Observations involving the same subject made under different conditions are very likely to be correlated to some degree. It is the presence of this correlation that makes the analysis of repeated measure data more complex than when each subject has only a single value of the response variable recorded.

There are a number of approaches to the analysis of repeated measures data. In this chapter, we consider an analysis of variance type model similar to those encountered in Chapters 5 and 6 that give valid results under a particular assumption about the pattern of correlations between the repeated measures, along with two possible procedures that can be applied when this assumption is thought to be invalid. In the next chapter, we look at some more general models for the analysis of repeated measures data, particularly useful when time is the only within-subject factor. Such data are then often labeled *longitudinal*.

A brief outline of the analysis of variance model appropriate for the repeated measures data in Table 7.1 is given in Box 7.1.

Box 7.1 ANOVA Approach to the Analysis of Repeated Measures

■ A suitable model for the observations in the reverse Stroop experiment is

$$y_{ijkl} = \mu + \alpha_j + \beta_k + \gamma_l + (\alpha\beta)_{jk} + (\alpha\gamma)_{jl} + (\beta\gamma)_{kl} +$$

$$(\alpha\beta\gamma)_{jkl} + u_i + \varepsilon_{ijkl}$$

where y_{ijkl} represent the observation for the ith subject in the lth group ($l = 1, 2$), under the jth type condition ($j = 1, 2$), and the kth cue condition ($k = 1, 2, 3$), $\alpha_j, \beta_k, \gamma_l$ represent the main effects of type, cue, and group, $(\alpha\beta)_{jk}, (\alpha\gamma)_{jl}, (\beta\gamma)_{kl}$ the first order interactions between these factors and $(\alpha\beta\gamma)_{jkl}$ the second order interaction. These effects are known as *fixed effects*. The u_i represents the effect of subject i and ε_{ijkl} the residual or error term. The u_i are assumed to have a normal distribution with mean zero and variance σ_u^2 and are termed *random effects*. The ε_{ijkl} are also assumed normally distributed with zero mean and variance σ^2.

■ The model is a simple example of what are known as *linear mixed effects model* (cf. Chapter 8).

■ This model leads to a partition of the total variation in the observations into parts due to between- and within-subject variation; each of these can then be further partitioned to generate *F*-tests for both within- and between-subject factors and their interactions. Details are given in Everitt (2001b).

■ For repeated measures data, the *F*-tests are valid under three assumptions about the data:

1. *Normality*: the data arise from populations with normal distribution.

2. *Homogeneity of variance*: the variances of the assumed normal distributions are equal.

3. *Sphericity*: the variances of the differences between all pairs of the repeated measurements are equal. This requirement implies that the covariances between pairs of repeated measures are equal and that the variances of each repeated measurement are also equal, i.e., the covariance matrix of the repeated measures must have the so-called *compound symmetry* pattern.

■ It is the last assumption that is most critical for assessing the within-subject factors in repeated measures data because if it does not hold, the *F*-tests used are positively biased leading to an increase in the Type I error. (Sphericity, of course, is not an issue when testing between-subject factors.)

■ When the assumption is not thought to be valid, then either the models to be described in the next chapter can be used (the preferred option), or one of the following two procedures associated with the analysis of variance approach applied.

a) Correction factors

Box (1954) and Greenhouse and Geisser (1959) considered the effects of departures from the sphericity assumption in the repeated measures analysis of variance. They demonstrated that the extent to which a set of repeated measures departs from the sphericity assumption can be summarized in terms of a parameter ε, which is a function of the variances and covariances of the repeated measures (see Everitt, 2001b, for an explicit definition). An estimate of this parameter based on the sample covariances and variances can be used to decrease the degrees of freedom of F-tests for the within-subjects effect to account for deviations from sphericity (in fact, two different estimates have been suggested; see text for details). In this way, larger mean square ratios will be needed to claim statistical significance than when the correction is not used, and thus the increased risk of falsely rejecting the null hypothesis is removed. The minimum (lower bound) value of the correction factor is $1/(p-1)$ where p is the number of repeated measures.

b) Multivariate analysis of variance

An alternative to the use of correction factors in the analysis of repeated measures data when the sphericity assumption is judged to be inappropriate is to use multivariate analysis of variance for constructed contrasts of the repeated measures that characterize each main effect and interaction (for details, see Everitt, 2001b). The advantage is that no assumptions are now made about the pattern of correlations between the repeated measurements. The variances are also allowed to vary between the repeated measures. A disadvantage of using MANOVA for repeated measures is often stated to be the technique's relatively low power when the assumption of compound symmetry is actually valid. However, Davidson (1972) shows that this is really only a problem with small sample sizes.

7.3 Analysis Using SPSS

In this experiment, it is of interest to assess whether performance on the Reverse Stroop Task varies with cognitive style. Cognitive style (with levels "field independent" and "field dependent") is a between-subject factor. Performance on the Stroop is measured by six repeated measures corresponding to the combinations of the levels of the two within-subject factors word type (two levels "color names" and "form names") and cue condition (three levels "normal," "congruent," and "incongruent"). Since it is possible that only specific aspects of the performance on the Stroop depend on cognitive style, we are interested in testing both, a main effect of cognitive style as well as interactions between cognitive style and factors word type and cue condition.

To convey this design to SPSS, the **Data View** spreadsheet has to be of the format indicated in Table 7.1. SPSS assumes that the rows of the data file correspond to independent measurement units (here the 24 subjects). Repeated outcome measures per such unit have to be represented by multiple columns. Here the spreadsheet includes six columns labeled fn, fc, fi, cn, cc, and ci containing response times, one for each combination of word type and cue condition. The between-subject factor, labeled **group**, is also included in the file as a further column. The within-subject factors will later be defined explicitly.

As always, it is good practice to "look" at the data before undertaking a formal analysis. Here, in order to get a first impression of the data, we generate box plots for each group and repeated measure using the instructions:

- Graphs – Boxplot...
- Check **Clustered** and **Summaries of separate variables** on the resulting **Boxplot** dialogue box.
- Move all six response variables into the **Boxes Represent** list of the next **Define Clustered Boxplots: Summaries of Separate Variables** dialogue box.
- Move the group variable (group) into the **Category Axis** list.
- Open the **Chart Editor** for the resulting graph.
- Insert a vertical line by double-clicking on the *x*-axis labels and checking **Grid lines** on the resulting **Category Axis** dialogue box.
- Edit the patterns by highlighting the respective box plots and using the **Format – Color...** and **Format – Fill Pattern** commands from the **Chart Editor** menu.

Display 7.1 shows the resulting box plots. As is often the case for reaction time data, we find that the distributions are positively skewed, particularly in the group with a field dependent cognitive style. The group also has greater variability than the others. We will, therefore, transform the data in order to achieve approximately symmetric distributions rather than relying on the robustness of ANOVA *F*-tests. Log-transformations are usually helpful for mapping positively skewed distributions with minimum value near one into symmetric ones and for stabilizing variances. The minimum observed response time was 134 milliseconds. We, therefore, use the **Compute** command (see Chapter 1, Display 1.13) to generate new transformed response times labeled ln_fn, ln_fc, ln_fi, ln_cn, ln_cc, and ln_ci according to the function $y = \ln(x - 134)$. A clustered box plot of the transformed response times is shown in Display 7.2 indicating that the transformation has been successful to some extent.

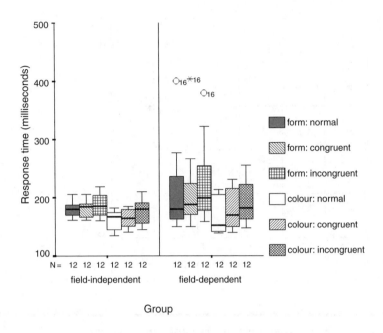

Display 7.1   Clustered box plot for Stroop response times.

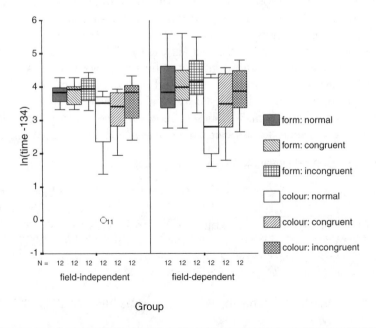

Display 7.2   Clustered box plot for log-transformed response times.

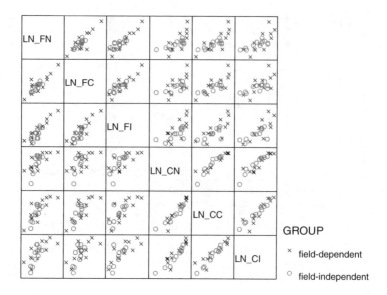

Display 7.3   Draughtman's plot of response times (log-scale).

A *draughtman's plot* provides another helpful look at the (transformed) data. This is a scatterplot matrix of the variables and enables us to assess the correlation structure between the repeated measures. We construct the scatterplot matrix as shown before (see Chapter 4, Display 4.12) and additionally include **group** under the **Set Markers by** list on the **Scatterplot Matrix** dialogue box. Display 7.3 shows the resulting scatterplot matrix. All pairs of repeated measures are positively correlated with the strength of the correlation appearing to be fairly constant and also similar across. The sample size is, however, rather small to judge the covariance structure accurately.

We now proceed to the repeated measures ANOVA using the commands

Analyze – General Linear Model – Repeated Measures...

This opens a dialogue box for declaring the within-subject factor(s) (Display 7.4). SPSS works on the assumption that each within-subject factor level (or factor level combination) corresponds to a single column in the **Data View** spreadsheet. This is the case for our data example; the six word type × condition combinations correspond to the columns labeled ln_fn, ln_fc, ln_fi, ln_cn, ln_cc, and ln_ci. We, therefore:

■ Declare the within-subject factors **wtype** and **cond** as shown in Display 7.4.
■ Select **Add** and **Define** to open the **Repeated Measures** dialogue box shown in Display 7.5.

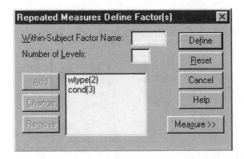

Display 7.4   Declaring the within-subject factors of a repeated measures design.

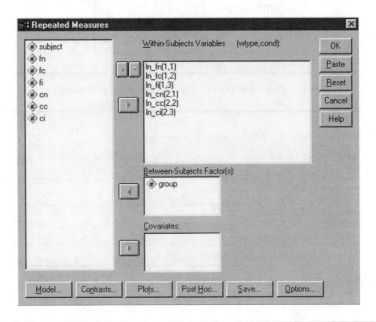

Display 7.5   Defining the levels of the within-subject factors and the between-subject factor(s).

- Insert the variables relating to each of the factor level combinations of the within-subject factors under the heading Within-Subjects Variables (wtype,cond).
- Insert the variable group under the Between-Subjects Factor(s) list.
- Check OK.

The resulting output is shown in Displays 7.6 to 7.10. SPSS automatically produces the results from both the ANOVA and MANOVA approaches to analyzing repeating measures. We will discuss each of these approaches in what follows.

Within-Subjects Factors

Measure: MEASURE_1

WTYPE	COND	Dependent Variable
1	1	LN_FN
	2	LN_FC
	3	LN_FI
2	1	LN_CN
	2	LN_CC
	3	LN_CI

Between-Subjects Factors

		Value Label	N
GROUP	1	field-indepe ndent	12
	2	field-depen dent	12

Display 7.6   Design summary output for Stroop response times.

Tests of Between-Subjects Effects

Measure: MEASURE_1

Transformed Variable: Average

Source	Type III Sum of Squares	df	Mean Square	F	Sig.
Intercept	1929.810	1	1929.810	728.527	.000
GROUP	1.927	1	1.927	.728	.403
Error	58.276	22	2.649		

Display 7.7   ANOVA table for testing the main effect of the between-subject factor on the Stroop response times.

SPSS supplies a design summary (Display 7.6). The "Within-Subjects Factors" table and the "Between-Subjects Factors" table repeat the definitions of the within-subject factors, wtype and cond, and the between-subject factor, group. (It should perhaps be noted that the ANOVA and MANOVA approaches to analyzing repeated measures will, in the presence of missing data, be based on the complete cases only.)

A single analysis is provided for testing the main effect(s) of and interactions between the between-subject factor(s). For each subject, SPSS simply calculates the average outcome over the repeated measures and then carries out an ANOVA of this new variable to test the effects of the between-subject factors (see previous two chapters). The one-way ANOVA

Tests of Within-Subjects Effects

Measure: MEASURE_1

Source		Type III Sum of Squares	df	Mean Square	F	Sig.
WTYPE	Sphericity Assumed	13.694	1	13.694	27.507	.000
	Greenhouse-Geisser	13.694	1.000	13.694	27.507	.000
	Huynh-Feldt	13.694	1.000	13.694	27.507	.000
	Lower-bound	13.694	1.000	13.694	27.507	.000
WTYPE * GROUP	Sphericity Assumed	.107	1	.107	.216	.647
	Greenhouse-Geisser	.107	1.000	.107	.216	.647
	Huynh-Feldt	.107	1.000	.107	.216	.647
	Lower-bound	.107	1.000	.107	.216	.647
Error(WTYPE)	Sphericity Assumed	10.952	22	.498		
	Greenhouse-Geisser	10.952	22.000	.498		
	Huynh-Feldt	10.952	22.000	.498		
	Lower-bound	10.952	22.000	.498		
COND	Sphericity Assumed	5.439	2	2.720	26.280	.000
	Greenhouse-Geisser	5.439	1.763	3.086	26.280	.000
	Huynh-Feldt	5.439	1.992	2.731	26.280	.000
	Lower-bound	5.439	1.000	5.439	26.280	.000
COND * GROUP	Sphericity Assumed	.126	2	6.307E-02	.609	.548
	Greenhouse-Geisser	.126	1.763	7.156E-02	.609	.529
	Huynh-Feldt	.126	1.992	6.333E-02	.609	.548
	Lower-bound	.126	1.000	.126	.609	.443
Error(COND)	Sphericity Assumed	4.553	44	.103		
	Greenhouse-Geisser	4.553	38.779	.117		
	Huynh-Feldt	4.553	43.815	.104		
	Lower-bound	4.553	22.000	.207		
WTYPE * COND	Sphericity Assumed	2.005	2	1.003	16.954	.000
	Greenhouse-Geisser	2.005	1.555	1.290	16.954	.000
	Huynh-Feldt	2.005	1.728	1.161	16.954	.000
	Lower-bound	2.005	1.000	2.005	16.954	.000
WTYPE * COND * GROUP	Sphericity Assumed	5.673E-03	2	2.836E-03	.048	.953
	Greenhouse-Geisser	5.673E-03	1.555	3.648E-03	.048	.918
	Huynh-Feldt	5.673E-03	1.728	3.283E-03	.048	.934
	Lower-bound	5.673E-03	1.000	5.673E-03	.048	.829
Error(WTYPE*COND)	Sphericity Assumed	2.602	44	5.914E-02		
	Greenhouse-Geisser	2.602	34.214	7.606E-02		
	Huynh-Feldt	2.602	38.012	6.846E-02		
	Lower-bound	2.602	22.000	.118		

Display 7.8   Univariate ANOVA tables for testing the main effects of and interactions involving the within-subject factors on the Stroop response times.

table for testing our single between-subject factor **group** is shown in Display 7.7. We do not find a significant main effect of the group factor ($F(1,22) = 0.73$, $p = 0.4$); averaged over the six Stroop tasks, response times do not differ between the two cognitive styles. (Note that such a test would not be of great interest if there were any interactions of **group** with the within-subject factors.) This ANOVA model assumes that the average (log transformed) response times have normal distributions with constant variances in each group; for our transformed data, these assumptions appear realistic.

Since the order of the Stroop reading tasks was randomized, it seems reasonable to assume sphericity on the grounds that any more complex covariance structure reflecting temporal similarities should have been

Mauchly's Test of Sphericity[b]

Measure: MEASURE_1

Within Subjects Effect	Mauchly's W	Approx. Chi-Square	df	Sig.	Epsilon[a]		
					Greenhous e-Geisser	Huynh-Feldt	Lower-bound
WTYPE	1.000	.000	0	.	1.000	1.000	1.000
COND	.865	3.036	2	.219	.881	.996	.500
WTYPE * COND	.714	7.075	2	.029	.778	.864	.500

Tests the null hypothesis that the error covariance matrix of the orthonormalized transformed dependent variables is proportional to an identity matrix.

a. May be used to adjust the degrees of freedom for the averaged tests of significance. Corrected tests are displayed in the Tests of Within-Subjects Effects table.

b.

Design: Intercept+GROUP
Within Subjects Design: WTYPE+COND+WTYPE*COND

Display 7.9   Mauchly's test for testing the sphericity assumption and correction factors.

Multivariate Tests [b]

Effect		Value	F	Hypothesis df	Error df	Sig.
WTYPE	Pillai's Trace	.556	27.507[a]	1.000	22.000	.000
	Wilks' Lambda	.444	27.507[a]	1.000	22.000	.000
	Hotelling's Trace	1.250	27.507[a]	1.000	22.000	.000
	Roy's Largest Root	1.250	27.507[a]	1.000	22.000	.000
WTYPE * GROUP	Pillai's Trace	.010	.216[a]	1.000	22.000	.647
	Wilks' Lambda	.990	.216[a]	1.000	22.000	.647
	Hotelling's Trace	.010	.216[a]	1.000	22.000	.647
	Roy's Largest Root	.010	.216[a]	1.000	22.000	.647
COND	Pillai's Trace	.668	21.141[a]	2.000	21.000	.000
	Wilks' Lambda	.332	21.141[a]	2.000	21.000	.000
	Hotelling's Trace	2.013	21.141[a]	2.000	21.000	.000
	Roy's Largest Root	2.013	21.141[a]	2.000	21.000	.000
COND * GROUP	Pillai's Trace	.043	.466[a]	2.000	21.000	.634
	Wilks' Lambda	.957	.466[a]	2.000	21.000	.634
	Hotelling's Trace	.044	.466[a]	2.000	21.000	.634
	Roy's Largest Root	.044	.466[a]	2.000	21.000	.634
WTYPE * COND	Pillai's Trace	.516	11.217[a]	2.000	21.000	.000
	Wilks' Lambda	.484	11.217[a]	2.000	21.000	.000
	Hotelling's Trace	1.068	11.217[a]	2.000	21.000	.000
	Roy's Largest Root	1.068	11.217[a]	2.000	21.000	.000
WTYPE * COND * GROUP	Pillai's Trace	.009	.098[a]	2.000	21.000	.907
	Wilks' Lambda	.991	.098[a]	2.000	21.000	.907
	Hotelling's Trace	.009	.098[a]	2.000	21.000	.907
	Roy's Largest Root	.009	.098[a]	2.000	21.000	.907

a. Exact statistic

b.

Design: Intercept+GROUP
Within Subjects Design: WTYPE+COND+WTYPE*COND

Display 7.10   Multivariate ANOVA output for testing the main effect and interactions involving the within-subject factors on the Stroop response times.

destroyed. The within-subjects parts of the linear mixed effects model ANOVA tables (for details, see Box 7.1) are included in the "Tests of Within-Subjects Effects" table under the rows labeled "Sphericity assumed" (Display 7.8).

For the construction of the relevant *F*-tests, repeated measures are aggregated in different ways before fitting ANOVA models. The first three rows of cells of the table relate to a model for the average (log-) response times for the two word types (that is averaged over the three cue conditions). Variability in these times can be due to a main effect of **wtype** or an interaction between **wtype** and the between-subject factor **group**. As a result, *F*-tests are constructed that compare variability due to these sources against within-subjects error variability in this model. We find a main effect of word type ($F(1,22) = 27.4$, $p < 0.001$) but no evidence of an interaction ($F(1,22) = 0.2$, $p = 0.65$). The next three rows of cells in the table relate to a model for the average response times for the three cue conditions. We again find a main effect of **cond** ($F(2,44) = 26.3$, $p < 0.001$) but no evidence for the two-way interaction involving **group** and **cond** ($F(2,44) = 0.6$, $p = 0.55$). Finally, the last three rows of cells relate to a model for the (log-) response time differences between color and form names for each cue condition. Since an interaction effect of the two within-subject factors implies that the differences between the levels of one factor depend on the level of the second factor, the last model allows assessing the interaction between **wtype** and **cond** and the three-way interaction involving all factors. There is evidence of an interaction effect of word type by cue condition ($F(2,44) = 17$, $p < 0.001$) but no evidence of a three-way interaction ($F(2,44) = 0.05$, $p = 0.95$).

The results involving the two within-subject factors discussed earlier rely on the assumption of sphericity. For these data, given that the order of presentation was randomized, such an assumption seems reasonable. In other situations, sphericity may be less easy to justify *a priori* (for example, if the within-subject order had *not* been randomized in this experiment), and investigators may then wish to test the condition formally for their data. This may be done using *Mauchly's test* (Krzanowski and Marriot, 1995). The test is automatically supplied by SPSS for each part of the within-subject factor analysis, except when a factor contains only two levels, in which case there is no correlation structure to consider (Display 7.9). Here the Mauchly test for average response times for cue conditions ($X^2(2) = 3$, $p = 0.22$) suggests that sphericity is an acceptable assumption. For differences in response times, however, the corresponding test ($X^2(2) = 7.1$, $p = 0.03$) sheds some doubt on the assumption.

When sphericity is in doubt, the part of the "Within-Subjects Effects" table in Display 7.8 corresponding to either of the correction terms might be used, or for a conservative approach, the lower bound rows could be employed. SPSS supplies estimates of the correction factors (one due to Greenhouse and Geisser, 1959, and one due to Huynh and Feldt, 1976) and the lower bounds together with Mauchly's tests (Display 7.9) and uses them to correct the degrees of freedom of the F-tests in the "Tests of Within-Subjects Effects" table (for details see Box 7.1). Here, the within-subject effects that tested significant under the assumption of sphericity remain highly significant even after correction (e.g., using Huynh Feldt correction factors, main effect cond: $F(1.99,43.82) = 26.3$, $p < 0.001$, inter-action effect wtype×cond: $F(1.73,38.01) = 17$, $p < 0.001$). However, on the whole, the correction factor approach is not recommended since other more acceptable approaches are now available (see Chapter 8).

Like the ANOVA approach, the MANOVA approach to repeated mea-sures also reduces the number of repeated measures before modeling contrasts of them by a multivariate ANOVA model (for details, see Everitt, 2001b). The results from fitting respective MANOVA models to test within-subject effects are shown in the "Multivariate Tests" table (Display 7.10). For each effect, SPSS displays the four commonly used multivariate tests (see Chapter 5). We again note that the test results for testing the main effect of wtype and the interaction wtype×group are identical to those obtained from the univariate ANOVA model in Display 7.8 since the tests are based on only two repeated measures (word type averages). The MANOVA approach does not alter the conclusions drawn about the remaining within-subject effects. As before, we find a significant main effect of cond (Wilk's lambda: $F(2,21) = 21.1$, $p < 0.001$) and evidence for an interaction between factors wtype and cond (Wilk's lambda: $F(2,21) = 11.2$, $p < 0.001$).

All the analyses reported above have established that the response times are affected by the two factors defining the Reverse Stroop test but not by cognitive style. It remains to interpret this interaction. As mentioned in Chapter 6 in the context of standard ANOVA, a line chart of estimated means is useful for this purpose. A line chart of predicted mean log-response times or "profile plot" can be generated via the **Plots...** button on the **Repeated Measures** dialogue box (see Display 7.5). Since the factor group is not involved in any significant main or interaction effects, we choose to display mean response times for each of the six repeated measures across groups. We display the levels of factor cond on the horizontal axis and use those of wtype to define separate lines. The resulting line chart is shown in Display 7.11, showing immediately the nature of our detected interaction. Participants were generally slower in reading form names than color names with the delay time greatest under normal cue conditions.

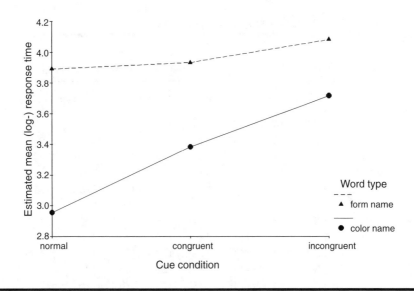

Display 7.11   Line chart of estimated mean (log-) response times.

7.4 Exercises

7.4.1 More on the Reverse Stroop Task

In the analysis of the Stroop data carried out in this chapter, a log transformation was used to overcome the data's apparent skewness. But interpreting results for log-reaction times may not be straightforward. A more appealing transformation may be to take the reciprocal of reaction time, making the dependent variable "speed." Reanalyze the data using this transformation and compare your results with those in the text.

7.4.2 Visual Acuity Data

The data in Table 7.2 arise from an experiment in which 20 subjects had their response times measured when a light was flashed into each of their eyes through lenses of powers 6/6, 6/18, 6/36, and 6/60. (A lens of power a/b means that the eye will perceive as being at a feet an object that is actually positioned at b feet). Measurements are in milliseconds. Here there are two within-subject factors.

Carry out an appropriate analysis to test whether the response times differ between the left and the right eye and whether such a difference depends on length strength.

Table 7.2   Visual Acuity and Lens Strength

Subject	Left Eye				Right Eye			
	6/6	6/18	6/36	6/60	6/6	6/18	6/36	6/60
1	116	119	116	124	120	117	114	122
2	110	110	114	115	106	112	110	110
3	117	118	120	120	120	120	120	124
4	112	116	115	113	115	116	116	119
5	113	114	114	118	114	117	116	112
6	119	115	94	116	100	99	94	97
7	110	110	105	118	105	105	115	115
8	117	119	119	123	110	110	115	111
9	116	120	123	124	115	115	120	114
10	118	118	119	120	112	120	120	119
11	120	119	118	121	110	119	118	122
12	110	112	115	118	120	123	123	124
13	118	120	122	124	120	118	116	122
14	118	120	120	122	112	120	122	124
15	110	110	110	120	120	119	118	124
16	112	112	114	118	110	110	108	120
17	112	112	120	122	105	110	115	105
18	105	105	110	110	118	120	118	120
19	110	120	122	123	117	118	118	122
20	105	120	122	124	100	105	106	110

7.4.3 Blood Glucose Levels

The data in Table 7.3 arise from a study in which blood glucose levels (mmol/liter) were recorded for six volunteers before and after they had eaten a test meal. Recordings were made at times −15, 0, 30, 60, 90, 120, 180, 240, 300, and 360 minutes after feeding time. The whole process was repeated four times, with the meals taken at various times of the day and night. Primary interest lies in assessing the time-of-day effect. Analyze the data using an appropriate repeated measures analysis of variance.

Table 7.3   Blood Glucose Levels

Subject	-15	0	30	60	90	120	180	240	300	360
10 am meal										
1	4.90	4.50	7.84	5.46	5.08	4.32	3.91	3.99	4.15	4.41
2	4.61	4.65	7.9	6.13	4.45	4.17	4.96	4.36	4.26	4.13
3	5.37	5.35	7.94	5.64	5.06	5.49	4.77	4.48	4.39	4.45
4	5.10	5.22	7.2	4.95	4.45	3.88	3.65	4.21	4.30	4.44
5	5.34	4.91	5.69	8.21	2.97	4.3	4.18	4.93	5.16	5.54
6	6.24	5.04	8.72	4.85	5.57	6.33	4.81	4.55	4.48	5.15
2 pm meal										
1	4.91	4.18	9.00	9.74	6.95	6.92	4.66	3.45	4.20	4.63
2	4.16	3.42	7.09	6.98	6.13	5.36	6.13	3.67	4.37	4.31
3	4.95	4.40	7.00	7.8	7.78	7.3	5.82	5.14	3.59	4.00
4	3.82	4.00	6.56	6.48	5.66	7.74	4.45	4.07	3.73	3.58
5	3.76	4.70	6.76	4.98	5.02	5.95	4.9	4.79	5.25	5.42
6	4.13	3.95	5.53	8.55	7.09	5.34	5.56	4.23	3.95	4.29
6 pm meal										
1	4.22	4.92	8.09	6.74	4.30	4.28	4.59	4.49	5.29	4.95
2	4.52	4.22	8.46	9.12	7.50	6.02	4.66	4.69	4.26	4.29
3	4.47	4.47	7.95	7.21	6.35	5.58	4.57	3.90	3.44	4.18
4	4.27	4.33	6.61	6.89	5.64	4.85	4.82	3.82	4.31	3.81
5	4.81	4.85	6.08	8.28	5.73	5.68	4.66	4.62	4.85	4.69
6	4.61	4.68	6.01	7.35	6.38	6.16	4.41	4.96	4.33	4.54
2 am meal										
1	4.05	3.78	8.71	7.12	6.17	4.22	4.31	3.15	3.64	3.88
2	3.94	4.14	7.82	8.68	6.22	5.1	5.16	4.38	4.22	4.27
3	4.19	4.22	7.45	8.07	6.84	6.86	4.79	3.87	3.6	4.92
4	4.31	4.45	7.34	6.75	7.55	6.42	5.75	4.56	4.30	3.92
5	4.3	4.71	7.44	7.08	6.3	6.5	4.5	4.36	4.83	4.5
6	4.45	4.12	7.14	5.68	6.07	5.96	5.20	4.83	4.50	4.71

Minutes after Meal

Chapter 8

Analysis of Repeated Measures II: Linear Mixed Effects Models; Computer Delivery of Cognitive Behavioral Therapy

8.1 Description of Data

The data to be used in this chapter arise from a clinical trial that is described in detail in Proudfoot et al. (2003). The trial was designed to assess the effectiveness of an interactive program using multi-media techniques for the delivery of cognitive behavioral therapy for depressed patients and known as *Beating the Blues* (BtB). In a randomized controlled trial of the program, patients with depression recruited in primary care were randomized to either the BtB program, or to *Treatment as Usual* (TAU). The outcome measure used in the trial was the *Beck Depression Inventory II* (Beck et al., 1996) with higher values indicating more depression. Measurements of this variable were made on five occasions:

- Prior to treatment
- Two months after treatment began
- At 1, 3, and 6 months follow-up, i.e., at 3, 5, and 8 months after treatment

The data for 100 patients, 48 in the TAU group and 52 in the BtB group, are shown in Table 8.1. These data are a subset of the original and are used with the kind permission of the organizers of the study, in particular Dr. Judy Proudfoot.

The data in Table 8.1 are repeated measures data with time as the single within-subject factor, i.e., they are longitudinal data. They have the following features that are fairly typical of those collected in many clinical trials in psychology and psychiatry:

- There are a considerable number of missing values caused by patients dropping out of the study (coded 999 in Table 8.1).
- There are repeated measurements of the outcome taken on each patient post treatment, along with a baseline pre-treatment measurement.
- There is interest in the effect of more than treatment group on the response; in this case, the effect of duration of illness categorized as whether a patient had been ill for longer than six months (duration = 1) or for less than six months (duration = 0).

The question of most interest about these data is whether the BtB program does better than TAU in treating depression.

8.2 Linear Mixed Effects Models

Longitudinal data were introduced as a special case of repeated measures in Chapter 7. Such data arise when subjects are measured on the same response variable on several different occasions. Because the repeated measures now arise solely from the passing of time, there is no possibility of randomizing the occasions, and it is this that essentially differentiates longitudinal data from other repeated measure situations arising in psychology and elsewhere. The special structure of longitudinal data make the sphericity condition described in Chapter 7 very difficult to justify. With longitudinal data, it is very unlikely that the measurements taken close to one another in time will have the same degree of correlation as measurements made at more widely spaced time intervals.

An analysis of longitudinal data requires application of a model that can represent adequately the average value of the response at any time

Table 8.1   A Subset of the Data from the Original BtB Trial

			Depression Score				
Subject	Duration[a]	Treatment[b]	Baseline	2 Months	3 Months	5 Months	8 Months
1	1	0	29	2	2	999[c]	999
2	1	1	32	16	24	17	20
3	0	0	25	20	999	999	999
4	1	1	21	17	16	10	9
5	1	1	26	23	999	999	999
6	0	1	7	0	0	0	0
7	0	0	17	7	7	3	7
8	1	0	20	20	21	19	13
9	0	1	18	13	14	20	11
10	1	1	20	5	5	8	12
11	1	0	30	32	24	12	2
12	0	1	49	35	999	999	999
13	1	0	26	27	23	999	999
14	1	0	30	26	36	27	22
15	1	1	23	13	13	12	23
16	0	0	16	13	3	2	0
17	1	1	30	30	29	999	999
18	0	1	13	8	8	7	6
19	1	0	37	30	33	31	22
20	0	1	35	12	10	8	10
21	1	1	21	6	999	999	999
22	0	0	26	17	17	20	12
23	1	0	29	22	10	999	999
24	1	0	20	21	999	999	999
25	1	0	33	23	999	999	999
26	1	1	19	12	13	999	999
27	0	0	12	15	999	999	999
28	1	0	47	36	49	34	999
29	1	1	36	6	0	0	2
30	0	1	10	8	6	3	3
31	0	0	27	7	15	16	0
32	0	1	18	10	10	6	8
33	0	1	11	8	3	2	15
34	0	1	6	7	999	999	999
35	1	1	44	24	20	29	14
36	0	0	38	38	999	999	999
37	0	0	21	14	20	1	8
38	1	0	34	17	8	9	13
39	0	1	9	7	1	999	999

Table 8.1 (continued)   A Subset of the Data from the Original BtB Trial

			Depression Score				
Subject	Duration	Treatment	Baseline	2 Months	3 Months	5 Months	8 Months
40	1	0	38	27	19	20	30
41	0	1	46	40	999	999	999
42	0	0	20	19	18	19	18
43	1	0	17	29	2	0	0
44	1	1	18	20	999	999	999
45	1	1	42	1	8	10	6
46	0	1	30	30	999	999	999
47	0	1	33	27	16	30	15
48	0	1	12	1	0	0	999
49	0	1	2	5	999	999	999
50	1	0	36	42	49	47	40
51	0	0	35	30	999	999	999
52	0	1	23	20	999	999	999
53	1	0	31	48	38	38	37
54	0	1	8	5	7	999	999
55	0	0	23	21	26	999	999
56	0	1	7	7	5	4	0
57	0	0	14	13	14	999	999
58	0	0	40	36	33	999	999
59	0	1	23	30	999	999	999
60	1	1	14	3	999	999	999
61	1	0	22	20	16	24	16
62	1	0	23	23	15	25	17
63	0	0	15	7	13	13	999
64	1	0	8	12	11	26	999
65	1	1	12	18	999	999	999
66	1	0	7	6	2	1	999
67	0	0	17	9	3	1	0
68	0	1	33	18	16	999	999
69	0	0	27	20	999	999	999
70	0	1	27	30	999	999	999
71	0	1	9	6	10	1	0
72	1	1	40	30	12	999	999
73	1	0	11	8	7	999	999
74	0	0	9	8	999	999	999
75	1	0	14	22	21	24	19
76	1	1	28	9	20	18	13
77	1	1	15	9	13	14	10
78	1	1	22	10	5	5	12

Table 8.1 (continued) A Subset of the Data from the Original BtB Trial

Subject	Duration	Treatment	Depression Score				
			Baseline	2 Months	3 Months	5 Months	8 Months
79	0	0	23	9	999	999	999
80	1	0	21	22	24	23	22
81	1	0	27	31	28	22	14
82	1	1	14	15	999	999	999
83	1	0	10	13	12	8	20
84	0	0	21	9	6	7	1
85	1	1	46	36	53	999	999
86	1	1	36	14	7	15	15
87	1	1	23	17	999	999	999
88	1	0	35	0	6	0	1
89	0	1	33	13	13	10	8
90	0	1	19	4	27	1	2
91	0	0	16	999	999	999	999
92	0	1	30	26	28	999	999
93	0	1	17	8	7	12	999
94	1	1	19	4	3	3	3
95	1	1	16	11	4	2	3
96	1	1	16	16	10	10	8
97	0	0	28	999	999	999	999
98	1	1	11	22	9	11	11
99	0	0	13	5	5	0	6
100	0	0	43	999	999	999	999

[a] Illness duration codes: "6 months or less" = 0, "more than 6 months" = 1
[b] Treatment codes: TAU = 0, BtB = 1
[c] 999 denotes a missing value

point in terms of covariates such as treatment group, time, baseline value, etc., and also account successfully for the observed pattern of dependences in those measurements. There are a number of powerful methods for analyzing longitudinal data that largely meet the requirements listed above. They are all essentially made up of two components. The first component consists of a regression model for the average response over time and the effects of covariates on this average response. The second component provides a model for the pattern of covariances or correlations between the repeated measures. Each component of the model involves a set of parameters that have to be estimated from the data. In most applications, it is the parameters reflecting the effects of covariates on the average response that will be of most interest. Although the parameters modeling the covariance structure of the observations will not, in general, be of prime interest (they are often regarded as so-called *nuisance parameters*), specifying the wrong model for the covariance structure can affect the results that *are* of concern. Diggle (1988), for example, suggests that overparameterization of the covariance model component (i.e., using too many parameters for this part of the model) and too restrictive a specification (too few parameters to do justice to the actual covariance structure in the data) may both invalidate inferences about the mean response profiles when the assumed covariance structure does not hold. Consequently, an investigator has to take seriously the need to investigate each of the two components of the chosen model. (The univariate analysis of variance approach to the analysis of repeated measures data described in Chapter 7 suffers from being too restrictive about the likely structure of the correlations in a longitudinal data set, and the multivariate option from overparameterization.)

Everitt and Pickles (2000) give full technical details of a variety of the models now available for the analysis of longitudinal data. Here we concentrate on one of these, *the linear mixed effects model* or *random effects model*. Parameters in these models are estimated by maximum likelihood or by a technique know as *restricted maximum likelihood*. Details of the latter and of how the two estimation procedures differ are given in Everitt and Pickles (2000).

Random effects models formalize the sensible idea that an individual's pattern of responses in a study is likely to depend on many characteristics of that individual, including some that are unobserved. These unobserved or unmeasured characteristics of the individuals in the study put them at varying predispositions for a positive or negative treatment response. The unobserved characteristics are then included in the model as random variables, i.e., random effects. The essential feature of a random effects model for longitudinal data is that there is natural heterogeneity across individuals in their responses over time and that this heterogeneity can be represented by an appropriate probability distribution. Correlation among observations from the same individual arises from sharing unobserved variables, for example, an increased propensity to the condition under investigation, or perhaps a predisposition to exaggerate symptoms. Conditional on the values of these random effects, the repeated measurements of the response variable are assumed to be independent, the so-called *local independence assumption*.

A number of simple random effects models are described briefly in Box 8.1.

Box 8.1 *Random Effects Models*

■ Suppose we have observations made in a longitudinal study at time points, t_1, t_2, ..., t_T.

■ Assume we have a single covariate, treatment group coded as a zero/one dummy variable.

■ We want to model the response at time t_j in terms of treatment group and time (we will assume there is no treatment × time interaction).

■ Two possibilities are the *random intercept* and *random intercept/random slope* models.

■ *Random intercept model*
  • Here the model for the response given by individual i at time t_j, y_{ij}, is modeled as

$$y_{ij} = \beta_0 + \beta_1 \text{ Group}_i + \beta_2 t_j + u_i + \varepsilon_{ij}$$

  where Group_i is the dummy variable indicating the group to which individual i belongs, β_0, β_1, and β_2 are the usual regression coefficients for the model; β_0 is the intercept, β_1 represents the treatment effect, and β_2 the slope of the linear regression of outcome on time.

  • The ε_{ij}s are the usual residual or "error" terms, assumed to be normally distributed with mean zero and variance σ^2.

  • The u_i terms are, in this case, random effects that model possible heterogeneity in the intercepts of the individuals, and are assumed normally distributed with zero mean and variance σ_u^2. (The ε_{ij} and u_i terms are assumed independent of one another.)

  • The random intercept model is illustrated graphically in Figure 8.1. Each individual's trend over time is parallel to the treatment group's average trend, but the intercepts differ. The repeated measurements of the outcome for an individual will vary about the individual's *own* regression line, rather than about the regression line for all individuals.

  • The presence of the u_i terms implies that the repeated measurements of the response have a particular pattern for their covariance matrix; specifically, the diagonal elements are each given by $\sigma^2 + \sigma_u^2$, and the off-diagonal elements are each equal to σ_u^2 — this is essentially the same as the analysis of variance model described in Chapter 7.

- As noted elsewhere in the chapter, the implication that each pair of repeated measurements has the same correlation is not a realistic one for most longitudinal data sets. In practice, it is more likely that observations made closer together in time will be more highly correlated than those taken further apart. Consequently, for many such data sets, the random intercept model will not do justice to the observed pattern of covariances between the repeated observations. A model that allows a more realistic structure for the covariances is one that allows heterogeneity in both slopes and intercepts.

■ *Random intercept and slope model*
- In this case the model is given by

$$y_{ij} = \beta_0 + \beta_1 \ \text{Group}_i + \beta_2 t_j + u_{i1} + u_{i2} t_j + \varepsilon_{ij}$$

- Here the u_{i1} terms model heterogeneity in intercepts and the u_{i2} terms, heterogeneity in slopes. The two random effects are generally assumed to have a *bivariate normal distribution* with zero means for both variables, variances, $\sigma^2_{u_1}$, $\sigma^2_{u_2}$ and covariance $\sigma_{u_1 u_2}$. This model is illustrated in Figure 8.2; individuals are allowed to deviate in terms of both slope and intercept from the average trend in their group. (In this chapter, when fitting the random intercept and slope model we will assume that the covariance term is zero.)
- This model allows a more complex pattern for the covariance matrix of the repeated measurements. In particular, it allows variances and covariances to change over time, a pattern that occurs in many longitudinal data sets. (An explicit formula for the covariance matrix implied by this model is given in Everitt, 2002b.)
- Tests of fit of competing models are available that allow the most appropriate random effects model for the data to be selected — again, see Everitt (2002b) for details of such tests. In practice, however, changing the random effects to be included in a model often does not alter greatly the estimates of the regression coefficient (or coefficients) associated with the fixed effect(s) (β_1 and β_2 in the two equations above) or their estimated standard errors.

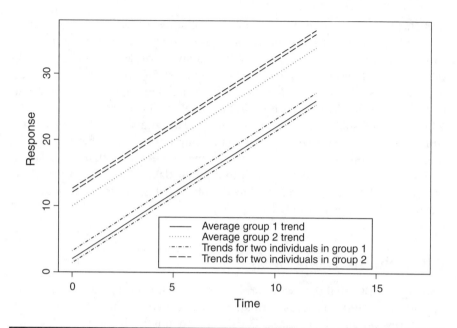

Figure 8.1   Graphical illustration of random intercept model.

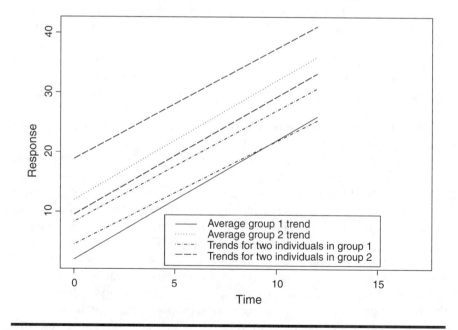

Figure 8.2   Graphical illustration of random intercept and slope model.

(The BtB data set contains a number of subjects who fail to attend all the scheduled post treatment visits — they *drop out* of the study. This can be a source of problems in the analysis of longitudinal data as described in Everitt and Pickles, 2000, who show that commonly used approaches such as *complete-case analysis* are flawed and can often lead to biased results. The random effects models to be used in this chapter, which, as we shall see, use all the *available* data on each subject, are valid under weaker assumptions than complete-case analysis although there are circumstances where they also may give results that are suspect. The whole issue of dealing with longitudinal data containing dropouts is complex and readers are referred to Everitt, 2002b, and Rabe-Hesketh and Skrondal, 2003, for full details.)

8.3 Analysis Using SPSS

Initially, the BtB trial data were organized in what is known as the standard or wide repeated measures format, i.e., as shown in Table 8.1. A row in the corresponding **Data View** spreadsheet contains both the repeated post treatment depression scores (at 2 months after start of treatment: **bdi2m**, 3 months: **bdi3m**, 5 months: **bdi5m**, and 8 months: **bdi8m**) for a trial participant and the values of other covariates for that individual, namely, baseline depression score (variable labeled **bdipre**), the treatment (**treat**), and illness duration factor (**duration**). A subject identifier (**subject**) is also included. For generating descriptive summaries or plots, it is useful to have the data set in this format, but as we will see later, for formal inferences using linear mixed effects models, we will need to convert the data set into what is often referred to as long format.

In order to get a first impression of the data, we generate a clustered error bar graph (corresponding to the treatment groups) for separate variables (i.e., the repeated measures). We choose error bars to represent confidence intervals. In addition, we open the **Options** sub-dialogue box from the **Define Clustered Error Bar...** dialogue box and check **Exclude cases variable by variable** so that summary statistics are based on the available data values for the variable in question (for more details on error bar charts, see Chapter 5, Display 5.8). Display 8.1 shows the error bar graph after some editing; indicating that the number of available cases with depression ratings decreases over time. The mean depression rating appears to decrease over time in both groups, with an apparent larger decrease in the BtB group during the initial two months post treatment period.

A plot of group means over time does not give any "feel" for the variability in individual response profiles and should be supplemented with a plot of individual temporal curves generated by using the commands

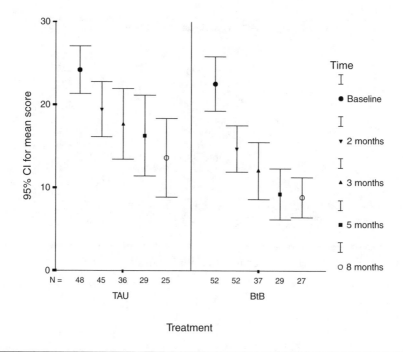

Display 8.1   Clustered error bar chart for BtB trial depression ratings.

Graphs – Line...

and then the following series of steps:

- Select **Multiple** and **Values of individual cases** in the resulting dialogue box.
- Fill in the next dialogue box as shown Display 8.2.
- Open the **Chart Editor** and used the commands **Series – Transpose Data** from the menu bar to swap the variables defining the lines and the *x*-axis categories.
- Edit further to improve appearance.

Since it is quite impossible to distinguish 100 separate lines on a single graph, we display subsets of cases at a time (by using **Data – Select Cases...** beforehand, see Chapter 1, Display 1.11). The line graph for the first ten cases is shown in Display 8.3. The graph illustrates the presence of *tracking* in the repeated measures data, i.e., subjects who start with high (low) tend to remain relatively high (low). Also apparent is the missing value pattern typical of dropout in longitudinal studies — once a subject withdraws from the study observations are missing for all subsequent time points.

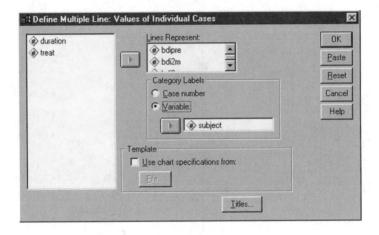

Display 8.2 Defining a line chart of multiple lines vs. a single variable.

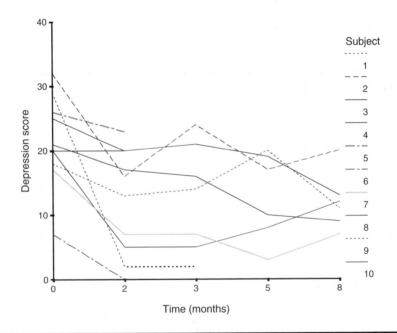

Display 8.3 Temporal depression ratings curves for the first ten subjects from Table 8.1.

Before proceeding to fit either of the random effects models described in Box 8.1 to the BtB data, it is important to examine the correlational structure of the depression values within each treatment group. This we can do by examining the appropriate correlation matrices. This will involve

a) TAU group

Correlations [a]

		Baseline depression score	Depression score at 2 months	Depression score at 3 months	Depression score at 5 months	Depression score at 8 months
Baseline depression score	Pearson Correlation	1	.613	.631	.502	.407
	Sig. (2-tailed)		.000	.000	.006	.044
	N	48	45	36	29	25
Depression score at 2 months	Pearson Correlation	.613	1	.818	.757	.716
	Sig. (2-tailed)	.000		.000	.000	.000
	N	45	45	36	29	25
Depression score at 3 months	Pearson Correlation	.631	.818	1	.863	.795
	Sig. (2-tailed)	.000	.000		.000	.000
	N	36	36	36	29	25
Depression score at 5 months	Pearson Correlation	.502	.757	.863	1	.850
	Sig. (2-tailed)	.006	.000	.000		.000
	N	29	29	29	29	25
Depression score at 8 months	Pearson Correlation	.407	.716	.795	.850	1
	Sig. (2-tailed)	.044	.000	.000	.000	
	N	25	25	25	25	25

a. Treatment = TAU

b) BtB group

Correlations [a]

		Baseline depression score	Depression score at 2 months	Depression score at 3 months	Depression score at 5 months	Depression score at 8 months
Baseline depression score	Pearson Correlation	1	.631	.538	.557	.414
	Sig. (2-tailed)		.000	.001	.002	.032
	N	52	52	37	29	27
Depression score at 2 months	Pearson Correlation	.631	1	.711	.763	.576
	Sig. (2-tailed)	.000		.000	.000	.002
	N	52	52	37	29	27
Depression score at 3 months	Pearson Correlation	.538	.711	1	.592	.402
	Sig. (2-tailed)	.001	.000		.001	.038
	N	37	37	37	29	27
Depression score at 5 months	Pearson Correlation	.557	.763	.592	1	.656
	Sig. (2-tailed)	.002	.000	.001		.000
	N	29	29	29	29	27
Depression score at 8 months	Pearson Correlation	.414	.576	.402	.656	1
	Sig. (2-tailed)	.032	.002	.038	.000	
	N	27	27	27	27	27

a. Treatment = BtB

Display 8.4 Within-group correlations between repeated measures.

use of the Data – Split file... commands (see Chapter 1, Display 1.10) followed by the steps outlined for Display 4.1 in Chapter 4. Here, we need to select Exclude cases pairwise in the Bivariate correlations: Options sub-dialogue box so all the available data for each participant in the trial will be used in the calculation of the required correlation coefficients. The resulting correlation matrices are shown in Display 8.4. The correlations of the repeated measures of depression are moderate to large and all positive. In the TAU group, there is an apparent decrease in the strength of the correlations with increasing length of time between the repeated measures, but this does not seem to be the case in the BtB group.

We are now nearly ready to fit our random effects models, but first the data need to be rearranged into long format in which each repeated

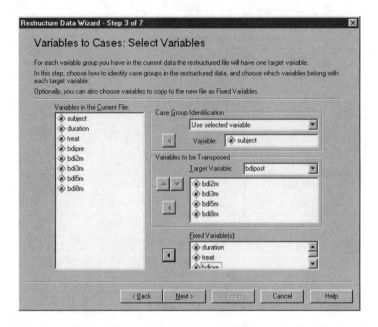

Display 8.5 Step 3 of the Restructure Data Wizard.

measurement and associated covariate value appears as a separate row in the data file. Repeated measures data can easily be converted from wide format into long format (or vice versa) by using SPSS's **Restructure Data Wizard**. The **Wizard** is called by the commands

Data – Restructure…

This starts up a set of dialogue boxes. For converting from wide into long format, the **Wizard** proceeds in seven steps:

1. The initial dialogue box contains Restructure selected variables into cases as the default option. We select Next to proceed to the next step.
2. The second dialogue box asks about the number of variable groups for restructuring. Here we need to rearrange one variable group, the post-treatment BDI variables, into one long combined variable. We therefore choose the default option One (for example, w1, w2, and w3) and click Next.
3. The dialogue box of step 3 is shown in Display 8.5. We choose our subject variable to identify cases in long format. We label the new variable to be constructed bdipost and indicate that the new variable is to be made up of the initial variables bdi2m, bdi3m, bdi5m,

	subject	duration	treat	bdipre	time	bdipost
1	1	1	0	29	1	2
2	1	1	0	29	2	2
3	1	1	0	29	3	999
4	1	1	0	29	4	999
5	2	1	1	32	1	16
6	2	1	1	32	2	24
7	2	1	1	32	3	17
8	2	1	1	32	4	20
9	3	0	0	25	1	20
10	3	0	0	25	2	999
11	3	0	0	25	3	999
12	3	0	0	25	4	999
13	4	1	1	21	1	17
14	4	1	1	21	2	16
15	4	1	1	21	3	10
16	4	1	1	21	4	9
17	5	1	1	26	1	23
18	5	1	1	26	2	999
19	5	1	1	26	3	999
20	5	1	1	26	4	999
21	6	0	1	7	1	0

Display 8.6 Part of Data View spreadsheet after restructuring into long format.

and bdi8m. The remaining variables duration, treat, and bdipre simply will have to be repeated four times to convert them into long format and this is requested by including them in the Fixed Variable(s) list.

4. On the next box where asked How many index variables do you want to create, we check the default option One since in a longitudinal study there is only one within subject factor (time) to be investigated.

5. In step 5, we choose the default option of giving sequential numbers to the index variable and relabel the index variable time.

6. When asked about the handling of the variables not selected, we keep the default option Drop variable(s) from the new data file since all our variables have previously been selected. When asked what to do with missing values in the transposed variables, we also choose the default option Create a case in the new file.

7. Finally, when asked What do you want to do?, we choose the default option Restructure the data now and click Finish.

This results in a rearrangement of our data file. The new long format data is illustrated in Display 8.6. The new Data View spreadsheet contains 400 rows for post-treatment depression scores of 100 cases at 4 time points. The new bdipost variable contains missing values (coded 999) with 97 out of the 100 cases contributing at least one post-treatment value.

We can now (finally) move on to fit a number of models to the BtB data beginning with one that makes the highly unrealistic assumption that the repeated measures of depression in the data are independent of one another. Our reason for starting with such an unpromising model is that it will provide a useful comparison with the two random effects models to be fitted later that both allow for nonindependence of these measures.

The independence model could be fitted using the SPSS multiple regression dialogue as described in Chapter 4, with post-treatment score as the dependent variable and time, treatment, baseline, and duration as explanatory variables (with the data in long form, this analysis will simply ignore that sets of four observations come from the same subject). But it is also possible to use the mixed models routine in SPSS to fit this independence model and this provides a convenient introduction to the procedure. Note that the categorical explanatory variables treatment (**treat**) and illness duration (**duration**) are represented by 0/1 variables in our **Data View** spreadsheet and so no further dummy-coding is necessary before regression modeling. The effect of time will be modeled by a linear effect of post-treatment month and, for this purpose, we generate a continuous variable **month** (with values 2, 3, 5, and 8) by recoding the **time** factor levels. For convenience, we further center this variable (new variable **c_month**) and the baseline BDI scores (new variable **c_pre**) by subtracting the respective sample means.

The mixed models routine is accessed by the commands

Analyze – Mixed Models – Linear...

The initial dialogue box titled **Linear Mixed Models: Specify Subjects and Repeated** is concerned with modeling the covariance structure of the data. Since for the moment we are assuming independence, we can simply proceed to the next dialogue box shown in Display 8.7. This **Linear Mixed Models** dialogue box serves to define the variables that we wish to include in the linear mixed effects model, and here we specify our dependent variable, **bdipost**, and our explanatory variables, **c_pre**, **duration**, **treat**, and **c_month**. Since our explanatory variables are either continuous or already dummy-coded, they can all be included under the **Covariate(s)** list.

Next, we need to specify which of the explanatory variables have fixed effects and which have random effects using the **Fixed...** and **Random...** buttons. Since, at the moment, we are dealing with a multiple regression model in which there are no random effects, we need only to employ the **Fixed....** button. Display 8.8 shows the resulting **Fixed Effects** sub-dialogue box where we specify fixed main effects for each of the four explanatory variables. The **Parameter estimates** and **Tests for Covariance parameters** options on the **Statistics** sub-dialogue box have also been checked to generate some further useful output.

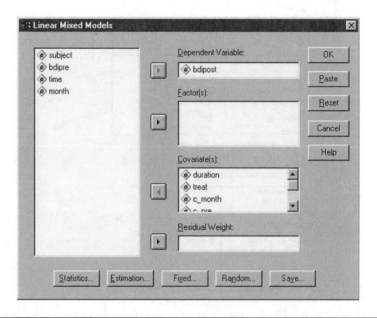

Display 8.7 Defining the variables to be used in a linear mixed effects model.

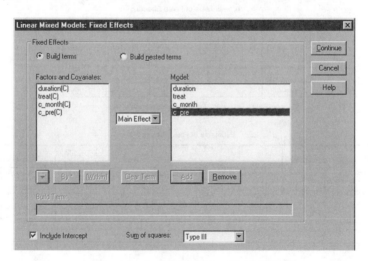

Display 8.8 Defining the fixed effects of a linear mixed effects model.

Display 8.9 shows the resulting output:

■ The "Model dimension" table simply summarizes the basics of the fitted model. In this case, the model includes an intercept parameter; four regression coefficients, one for each of the explanatory

Model Dimension [a]

		Number of Levels	Number of Parameters
Fixed Effects	Intercept	1	1
	DURATION	1	1
	TREAT	1	1
	C_MONTH	1	1
	C_PRE	1	1
Residual			1
Total		5	6

a. Dependent Variable: Post-treatment depression score .

Information Criteria [a]

-2 Restricted Log Likelihood	2008.634
Akaike's Information Criterion (AIC)	2010.634
Hurvich and Tsai's Criterion (AICC)	2010.649
Bozdogan's Criterion (CAIC)	2015.251
Schwarz's Bayesian Criterion (BIC)	2014.251

The information criteria are displayed in smaller-is-better forms.

a. Dependent Variable: Post-treatment depression score .

Type III Tests of Fixed Effects [a]

Source	Numerator df	Denominator df	F	Sig.
Intercept	1	275	217.151	.000
DURATION	1	275	3.983	.047
TREAT	1	275	17.321	.000
C_MONTH	1	275	16.188	.000
C_PRE	1	275	96.289	.000

a. Dependent Variable: Post-treatment depression score .

Estimates of Fixed Effects [a]

Parameter	Estimate	Std. Error	df	t	Sig.	95% Confidence Interval	
						Lower Bound	Upper Bound
Intercept	15. 15442	1.0283921	275	14. 736	.000	13.1299012	17.1789439
DURATION	2.2257989	1.1153411	275	1.996	.047	3.010731E-02	4.4214904
TREAT	-4.41121	1.0599089	275	-4.162	.000	-6.4977746	-2.3246421
C_MONTH	-.9657631	.2400315	275	-4.023	.000	-1.4382958	-.4932304
C_PRE	.5234639	5.33E-02	275	9.813	.000	.4184464	.6284814

a. Dependent Variable: Post-treatment depression score .

Estimates of Covariance Parameters [a]

Parameter	Estimate	Std. Error	Wald Z	Sig.	95% Confidence Interval	
					Lower Bound	Upper Bound
Residual	77.60373	6.6180686	11.726	.000	65.6586425	91.7219591

a. Dependent Variable: Post-treatment depression score .

Display 8.9   **Regression model output for BtB data.**

variables; and the variance parameter of the residual terms in the model.

▪ The "Information Criteria" table is helpful for comparing different random effects models, as we will see later.

▪ The "Type III Tests of Fixed Effects" and the "Estimates of Fixed Effects" tables give the F-tests and associated p-values for assessing the regression coefficients. For a model including only fixed effects as here, these tests (and the parameter estimates and standard errors that follow) are the same as those generated by the regression or ANCOVA routines in Chapters 4, 5, and 6. The results suggest that each of the four explanatory variables is predictive of post-treatment BDI. However, we shall not spend time interpreting these results in any detail since they have been derived under the unrealistic assumption that the repeated measurements of the BDI variable are independent.

▪ The "Estimates of Covariance parameters" table gives details of the single covariance parameter in our multiple regression model, namely, the variance of the error term. The table provides an estimate, confidence interval, and test for zero variance.

(Note that in this analysis, all the observations that a patient actually has are retained, unlike what would have happened if we had carried out a repeated measures (M)ANOVA on the basis of complete cases only as described in Chapter 7.)

Now we shall fit some random effects models proper to the BtB data, beginning with the random intercept model described in Box 8.1. This model allows a particular correlation structure for the repeated measures (see Box 8.1). To specify this model in SPSS, we need to go through the following steps:

▪ Add the subject identifier (**subject**) under the **Subjects** list of the initial **Linear Mixed Models: Specify Subjects and Repeated** dialogue box (Display 8.10).

▪ Include the variable **subject** under the **Factor(s)** list of the subsequent **Linear Mixed Models** dialogue box (see Display 8.7).

▪ Define the fixed effects as before using the **Fixed...** button (see Display 8.8), but this time, in addition, employ the **Random...** button to define the random effects part of the model.

▪ The random intercepts for subjects are specified in the **Random Effects** sub-dialogue box as illustrated in Display 8.11; the **subject** identifier is included under the **Combinations** list and **Include Intercept** is checked.

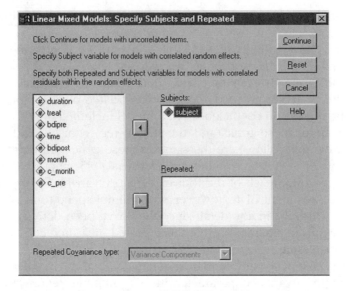

Display 8.10   Defining sets of correlated observations for subjects.

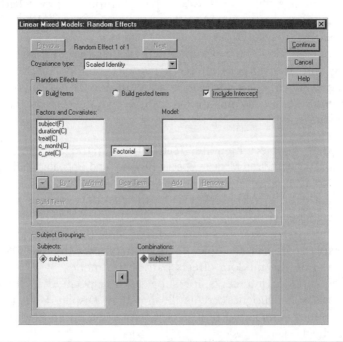

Display 8.11   Defining random intercepts for subjects.

Model Dimension[a]

		Number of Levels	Covariance Structure	Number of Parameters	Subject Variables
Fixed Effects	Intercept	1		1	
	DURATION	1		1	
	TREAT	1		1	
	C_MONTH	1		1	
	C_PRE	1		1	
Random Effects	Intercept	1	Identity	1	SUBJECT
Residual				1	
Total		6		7	

a. Dependent Variable: Post-treatment depression score.

Information Criteria[a]

-2 Restricted Log Likelihood	1872.390
Akaike's Information Criterion (AIC)	1876.390
Hurvich and Tsai's Criterion (AICC)	1876.434
Bozdogan's Criterion (CAIC)	1885.624
Schwarz's Bayesian Criterion (BIC)	1883.624

The information criteria are displayed in smaller-is-better forms.

a. Dependent Variable: Post-treatment depression score

Type III Tests of Fixed Effects[a]

Source	Numerator df	Denominator df	F	Sig.
Intercept	1	93.623	114.393	.000
DURATION	1	91.880	.124	.725
TREAT	1	89.637	3.784	.055
C_MONTH	1	193.921	23.065	.000
C_PRE	1	93.232	60.695	.000

a. Dependent Variable: Post-treatment depression score.

Display 8.12   Random intercept model output for BtB data.

The results from fitting the random intercepts model are shown in Display 8.12. The "Model Dimension" table indicates that subject random effects for the intercept have been added to the model. This results in fitting one extra model parameter, the variance of the subject random effects. We can use the first term given in the "Information Criteria" table to compare the random intercept model with the previously fitted independence model (see Display 8.9). The difference in $-2 \times$ Restricted Log Likelihood for the two models can be tested as a chi-square with degrees of freedom given by the difference in the number of parameters in the two models (this is known formally as a *likelihood ratio test*). Here the difference in the two likelihoods is $2008.634 - 1872.39 = 136.244$, which tested as a chi-square

Estimates of Fixed Effects [a]

Parameter	Estimate	Std. Error	df	t	Sig.	95% Confidence Interval	
						Lower Bound	Upper Bound
Intercept	16.36947	.5305030	93.623	10.695	.000	13.3304605	19.4084784
DURATION	.5910694	.6766526	91.880	.353	.725	-2.7389655	3.9211043
TREAT	-3.18832	.6389947	89.637	-1.945	.055	-6.4446523	.800538E-02
C_MONTH	.7052452	.1468458	193.921	-4.803	.000	-.9948652	-.4156253
C_PRE	.6107708	7.84E-02	93.232	7.791	.000	.4550940	.7664476

a. Dependent Variable: Post-treatment depression score.

Estimates of Covariance Parameters [a]

Parameter	Estimate	Std. Error	Wald Z	Sig.	95% Confidence Interval	
					Lower Bound	Upper Bound
Residual	25.24724	.6240408	9.62 2	.000	20.5942157	30.9515532
Intercept [subject ID diagonal SC]	53.17547	.3406957	5.69 3	.000	37.6869546	75.0294403

a. Dependent Variable: Post-treatment depression score.

Display 8.12 (continued)

with one degree of freedom, has an associated p-value less than 0.001. The random intercept model clearly provides a better fit for the BtB data than the independence model. The variance of the subject intercept effect is clearly not zero as confirmed by the confidence interval for the parameter, [37.7, 75], given later in Display 8.12 (see table "Estimates of Covariance Parameters"), which also gives an alternate test, the Wald test, for the null hypothesis that the variance parameter is zero (see Everitt and Pickles, 2000, for details). (Note that the p-value of the likelihood ratio test can be calculated easily within SPSS by using the SIG.CHISQ(q,df) function which returns the p-value of a test when the test statistic q has a chi-squared distribution with df degrees of freedom under the null hypothesis.)

The other terms given in the "Information Criteria" part of the output can also be used to compare models, see Everitt (2002b) for details.

The two output tables dealing with the fixed effects of the model show that the standard errors of the effects of the explanatory variables that explain between subject variability, namely **duration, treat,** and **c_pre,** have been increased by about 50% compared to the regression model. In contrast, the standard error of the effect of variable **c_month,** which explains within-subject variability, has been reduced by about 40%. This is intuitively what might be expected since for between-subject effects the correlations between repeated measures will lower the effective sample size compared to the same number of independent observations, and calculation of variability ignoring that measurements are made within subjects will lead to values that are too high. The result of these changes is that both **duration** and **treat** are not now significant at the 5% level.

The random intercept model implies a particular structure for the covariance matrix of the repeated measures, namely, that variances at the different post-treatment time points are equal, and covariances between each pair of time points are equal. In general, these are rather restrictive and unrealistic assumptions; for example, covariances between observations made closer together in time are likely to be higher than those made at greater time intervals and there was a suggestion that this might be the case for our data (Display 8.4). Consequently, a random intercept and slope model (as described in Box 8.1) that allows a less restrictive covariance structure for the repeated measures (the explicit structure is given in Everitt and Dunn, 2001) might provide an improvement in fit over the random intercept model.

In a random intercept and slope model, both the intercept term and the regression coefficient of BDI on month are allowed to vary between subjects. To introduce the random slope term, we again employ the Random... button. Random coefficients of the time variable c_month are specified in the Random Effects sub-dialogue box by adding a main effect of c_month to the Model list (see Display 8.11). SPSS understands this as an instruction to fit a slope for each subject, or in other words, a subject by c_month interaction, since the identifier subject is listed under the Combinations list. In fact, an alternative way of specifying the random slope and intercept model would be not to use the Combinations and Include Intercept facilities, but instead include a main effect of subject and a subject × c_month interaction under the Model list. (Note that at this stage we also removed the fixed main effect of duration from the model by excluding this variable from the Model list of the Fixed Effects sub-dialogue box, see Display 8.8.)

Fitting the random intercept and slope model initially results in SPSS displaying a warning about a "nonpositive definite Hessian matrix" (Display 8.13). This indicates that the fitting algorithm has not been able to produce an acceptable solution. When such a warning, or one relating to convergence problems, appears, subsequent output is invalid and should *not* be interpreted. The solution is generally to alter the default settings of the algorithm used to find the maximum likelihood solution. These settings can be accessed via the Estimation... button on the Linear Mixed Models dialogue box (see Display 8.7). Warnings about the Hessian matrix can often be dealt with by increasing the scoring option, so we

Warnings

> The final Hessian matrix is not positive definite although all convergence criteria are satisfied. The MIXED procedure continues despite this warning. Validity of subsequent results cannot be ascertained.

Display 8.13 SPSS warning indicating an unsuccessful optimization.

Display 8.14 Controlling the settings of the maximum likelihood algorithm.

change the Maximum scoring steps to 5 (Display 8.14). When nonconvergence occurs and the Hessian matrix is not nonpositive definite, increasing the number of iterations or reducing the accuracy of the solution by increasing the parameter convergence value might resolve the problem. In any case, whenever a warning is displayed, it is useful to have a look at the iteration history and this can be requested by checking Print iteration history on the Estimation sub-dialogue box (Display 8.14).

Changing the scoring option successfully resolves the problem on this occasion and Display 8.15 shows part of the output from fitting a random intercept and slope model. (Note that if increasing the scoring steps does not help to overcome a nonpositive definite Hessian matrix, it will probably be because of negative variance estimates or redundant effects.) The "Iteration History" table shows that all the parameter estimates of the random intercept and slope model have converged.

Model Dimension[a]

		Number of Levels	Covariance Structure	Number of Parameters	Subject Variables
Fixed Effects	Intercept	1		1	
	TREAT	1		1	
	C_MONTH	1		1	
	C_PRE	1		1	
Random Effects	Intercept	1	Identity	1	SUBJECT
	C_MONTH	1	Identity	1	SUBJECT
Residual				1	
Total		6		7	

a. Dependent Variable: Post-treatment depression score.

Iteration History [b]

					Covariance Parameters			Fixed Effect Parameters			
Iteration	Update Type	Number of Step-halvings	-2 Restricted Log Likelihood	Residual	Intercept [subject = SUBJECT] ID diagonal	C_MONTH [subject = SUBJECT] ID diagonal	Intercept	TREAT	C_MONTH	C_PRE
0	Initial	0	2021.046	26.14744	26.1474442	26.1474442	16.67599	-5.27073	1.1903886	.5125490
1	Scoring	1	1964.729	25.43479	41.2184411	2.8328789	16.63831	-4.50924	1.0287282	.5522275
2	Scoring	1	1926.724	24.90006	49.2784441	6.1882973	16.65211	-3.94426	-.8923691	.5796988
3	Scoring	1	1900.683	24.52781	52.9800480	3.0078202	16.67899	-3.59681	-.8017594	.5966415
4	Scoring	0	1875.382	24.04293	55.3249894	3.865E-02	16.71171	-3.20019	-.7000578	.6172862
5	Scoring	0	1874.882	23.96467	52.7822184	.2091313	16.70738	-3.23882	-.7087159	.6153676
6	Newton	0	1874.880	23.96393	53.0260027	.2012500	16.70778	-3.23618	-.7081400	.6154884
7	Newton	0	1874.880	23.96392	53.0277710	.2013409	16.70778	-3.23619	-.7081412	.6154881
8	Newton	4	1874.880 [a]	23.96392	53.0277710	.2013409	16.70778	-3.23619	-.7081412	.6154881

a. All convergence criteria are satisfied.

b. Dependent Variable: Post-treatment depression score .

Estimates of Covariance Parameters [a]

					95% Confidence Interval	
Parameter	Estimate	Std. Error	Wald Z	Sig.	Lower Bound	Upper Bound
Residual	23.96392	.9905226	8.013	.000	18.7643740	30.6042397
Intercept [subject = ID diagonal SC]	53.02777	.2225126	5.750	.000	37.7106173	74.5663874
C_MONTH [subject ID diagonal SC]	.2013409	.2975969	.677	.499	.111200E-02	3.6481403

a. Dependent Variable: Post-treatment depression score .

Display 8.15 Part of random intercept and slope model output for BtB data.

The Wald test (Wald $z = 0.68$, $p = 0.5$) demonstrates that the variance of the random slope effects is not significantly different from zero and here the simpler random intercept model provides an adequate description of the data. This might be because there is little departure of the observed correlation matrices in each treatment group from the compound symmetry structure allowed by the random intercept model, but it could also be a result of the random slope model we used being too restrictive since it did not allow for a possible correlation between the random intercept and random slope terms. SPSS version 11.0.1 (the one used in this text) does not allow for such a correlation, although version 11.5 and higher will allow this correlation to be a parameter in the model.

The final model chosen to describe the average profiles of BDI scores in the two groups over the four post-treatment visits is one that includes

Estimates of Fixed Effects[a]

Parameter	Estimate	Std. Error	df	t	Sig.	95% Confidence Interval	
						Lower Bound	Upper Bound
Intercept	16.70543	1.1920177	90.623	14.014	.000	14.3375015	19.0733606
TREAT	-3.22321	1.6303743	90.835	-1.977	.051	-6.4618315	1.540319E-02
C_MONTH	-.7039984	.1466745	194.693	-4.800	.000	-.9932734	-.4147235
C_PRE	.6165315	7.63E-02	93.058	8.076	.000	.4649276	.7681355

a. Dependent Variable: Post-treatment depression score .

Estimates of Covariance Parameters[a]

Parameter	Estimate	Std. Error	Wald Z	Sig.	95% Confidence Interval	
					Lower Bound	Upper Bound
Residual	25.22031	2.6180353	9.633	.000	20.5773777	30.9108366
Intercept [subject ID diagonal SUBJECT]	52.72001	9.2000578	5.730	.000	37.4484759	74.2193070

a. Dependent Variable: Post-treatment depression score .

Display 8.16   Parameter estimates of the final model for post-treatment BDI.

random intercepts for subjects and fixed effects of month, baseline BDI score and treatment (although this had a p-value slightly above the 5% level). The parameter estimates for this model are shown in Display 8.16. Subject heterogeneity accounts for the majority of the residual variability (estimated *intra-class correlation* 52.72/(52.72 + 25.22) × 100 = 67.6%). Conditional on treatment and visit mean post-treatment, BDI is estimated to increase by 0.62 points for each additional baseline BDI score (95% CI from 0.46 to 0.77 points) and conditional on pre-treatment score and treatment group to decrease 0.7 points each month (CI from 0.41 to 0.99 points). Finally, for a given visit and baseline BDI score the BtB treatment is estimated to decrease the mean BDI score by 3.22 points (CI from 0.015 points increase to 6.5 points decrease).

There are a number of residual diagnostics that can be used to check the assumptions of a linear mixed effects model. Here we demonstrate how to assess normality of the random effect terms. Checking out the conditional independence and homogeneity of variance is left as an exercise for the reader (Exercise 8.4.3).

The **Save...** button on the **Linear Mixed Effects** dialogue box allows the user to save predicted values and residuals (see Display 8.7). We check all three available options in the **Save** sub-dialogue box. This results in three new variables being added to the **Data View** spreadsheet after model fitting: participants' BDI scores predicted on the basis of fixed effects only (fxpred_1), values predicted after including the participants' estimated random effects (pred_1), and estimates of the errors (resid_1). From these we can calculate estimates of the subject random effects using the formula

a) Estimates of error term (residuals)

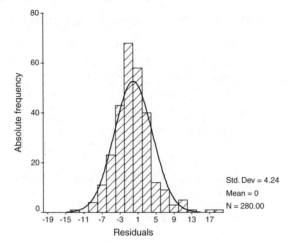

b) Estimates of random effects for subjects

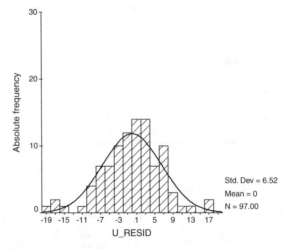

Display 8.17 Histograms for estimates of random effects.

u_resid = pred_1 − fxpred_1. We then construct histograms of the variables resid_1 and u_resid to check normality. (Before constructing the graph for the u_resid variable, we restrict the data set to time = 1 using the Select Cases... command to ensure that only one random effect per subject was used.) The resulting histograms do not indicate any departure from the normality assumption (Display 8.17).

8.4 Exercises

8.4.1 Salsolinol Levels and Alcohol Dependency

Two groups of subjects, one with moderate and the other with severe
dependence on alcohol, had their salsolinol secretion levels measured
(in mmol) on four consecutive days. The data are given in Table 8.2.
Primary interest is in whether the groups evolved differently over time.
Investigate this question using linear mixed effects models. (The raw data
are clearly skewed, so some kind of transformation will be needed).

8.4.2 Estrogen Treatment for Post-Natal Depression

The data in Table 8.3 arise from a double-blind, placebo controlled clinical
trial of a treatment for post-natal depression. The main outcome variable
was a composite measure of depression recorded on two occasions before
randomization to treatment and on six two-monthly visits after randomization. Not all the women in the trial had the depression variable recorded
on all eight scheduled visits. Use appropriate random effects models to
investigate these data, in particular to determine whether the active treatment helps reduce post-natal depression.

**Table 8.2 Salsolinol Secretion Levels of Alcohol-
Dependent Subjects**

	Day			
	1	*2*	*3*	*4*
Group 1 (Moderate)	0.33	0.70	2.33	3.20
	5.30	0.90	1.80	0.70
	2.50	2.10	1.12	1.01
	0.98	0.32	3.91	0.66
	0.39	0.69	0.73	2.45
	0.31	6.34	0.63	3.86
Group 2 (Severe)	0.64	0.70	1.00	1.40
	0.73	1.85	3.60	2.60
	0.70	4.20	7.30	5.40
	0.40	1.60	1.40	7.10
	2.60	1.30	0.70	0.70
	7.80	1.20	2.60	1.80
	1.90	1.30	4.40	2.80
	0.50	0.40	1.10	8.10

Table 8.3 Depression Ratings for Sufferers of Post-Natal Depression

| Subject | Group[a] | Before Randomization | | After Randomization | | | | | |
		Time 1	Time 2	Time 3	Time 4	Time 5	Time 6	Time 7	Time 8
1	0	18.00	18.00	17.00	18.00	15.00	17.00	14.00	15.00
2	0	25.11	27.00	26.00	23.00	18.00	17.00	12.00	10.00
3	0	19.00	16.00	17.00	14.00	n.a.	n.a.	n.a.	n.a.
4	0	24.00	17.00	14.00	23.00	17.00	13.00	12.00	12.00
5	0	19.08	15.00	12.00	10.00	8.00	4.00	5.00	5.00
6	0	22.00	20.00	19.00	11.54	9.00	8.00	6.82	5.05
7	0	28.00	16.00	13.00	13.00	9.00	7.00	8.00	7.00
8	0	24.00	28.00	26.00	27.00	n.a.	n.a.	n.a.	n.a.
9	0	27.00	28.00	26.00	24.00	19.00	13.94	11.00	9.00
10	0	18.00	25.00	9.00	12.00	15.00	12.00	13.00	20.00
11	0	23.00	24.00	14.00	n.a.	n.a.	n.a.	n.a.	n.a.
12	0	21.00	16.00	19.00	13.00	14.00	23.00	15.00	11.00
13	0	23.00	26.00	13.00	22.00	n.a.	n.a.	n.a.	n.a.
14	0	21.00	21.00	7.00	13.00	n.a.	n.a.	n.a.	n.a.
15	0	22.00	21.00	18.00	n.a.	n.a.	n.a.	n.a.	n.a.
16	0	23.00	22.00	18.00	n.a.	n.a.	n.a.	n.a.	n.a.
17	0	26.00	26.00	19.00	13.00	22.00	12.00	18.00	13.00
18	0	20.00	19.00	19.00	7.00	8.00	2.00	5.00	6.00
19	0	20.00	22.00	20.00	15.00	20.00	17.00	15.00	13.73
20	0	15.00	16.00	7.00	8.00	12.00	10.00	10.00	12.00
21	0	22.00	21.00	19.00	18.00	16.00	13.00	16.00	15.00
22	0	24.00	20.00	16.00	21.00	17.00	21.00	16.00	18.00
23	0	n.a.	17.00	15.00	n.a.	n.a.	n.a.	n.a.	n.a.
24	0	24.00	22.00	20.00	21.00	17.00	14.00	14.00	10.00
25	0	24.00	19.00	16.00	19.00	n.a.	n.a.	n.a.	n.a.
26	0	22.00	21.00	7.00	4.00	4.19	4.73	3.03	3.45
27	0	16.00	18.00	19.00	n.a.	n.a.	n.a.	n.a.	n.a.
28	1	21.00	21.00	13.00	12.00	9.00	9.00	13.00	6.00
29	1	27.00	27.00	8.00	17.00	15.00	7.00	5.00	7.00
30	1	24.00	15.00	8.00	12.27	10.00	10.00	6.00	5.96
31	1	28.00	24.00	14.00	14.00	13.00	12.00	18.00	15.00
32	1	19.00	15.00	15.00	16.00	11.00	14.00	12.00	8.00
33	1	17.00	17.00	9.00	5.00	3.00	6.00	.00	2.00
34	1	21.00	20.00	7.00	7.00	7.00	12.00	9.00	6.00
35	1	18.00	18.00	8.00	1.00	1.00	2.00	.00	1.00
36	1	24.00	28.00	11.00	7.00	3.00	2.00	2.00	2.00
37	1	21.00	21.00	7.00	8.00	6.00	6.50	4.64	4.97
38	1	19.00	18.00	8.00	6.00	4.00	11.00	7.00	6.00
39	1	28.00	27.46	22.00	27.00	24.00	22.00	24.00	23.00

**Table 8.3 (continued)   Depression Ratings for Sufferers
of Post-Natal Depression**

Subject	Group[a]	Before Randomization		After Randomization					
		Time 1	Time 2	Time 3	Time 4	Time 5	Time 6	Time 7	Time 8
40	1	23.00	19.00	14.00	12.00	15.00	12.00	9.00	6.00
41	1	21.00	20.00	13.00	10.00	7.00	9.00	11.00	11.00
42	1	18.00	16.00	17.00	26.00	n.a.	n.a.	n.a.	n.a.
43	1	22.61	21.00	19.00	9.00	9.00	12.00	5.00	7.00
44	1	24.24	23.00	11.00	7.00	5.00	8.00	2.00	3.00
45	1	23.00	23.00	16.00	13.00	n.a.	n.a.	n.a.	n.a.
46	1	24.84	24.00	16.00	15.00	11.00	11.00	11.00	11.00
47	1	25.00	25.00	20.00	18.00	16.00	9.00	10.00	6.00
48	1	n.a.	28.00	n.a.	n.a.	n.a.	n.a.	n.a.	n.a.
49	1	15.00	22.00	15.00	17.57	12.00	9.00	8.00	6.50
50	1	26.00	20.00	7.00	2.00	1.00	.00	.00	2.00
51	1	22.00	20.00	12.13	8.00	6.00	3.00	2.00	3.00
52	1	24.00	25.00	15.00	24.00	18.00	15.19	13.00	12.32
53	1	22.00	18.00	17.00	6.00	2.00	2.00	.00	1.00
54	1	27.00	26.00	1.00	18.00	10.00	13.00	12.00	10.00
55	1	22.00	20.00	27.00	13.00	9.00	8.00	4.00	5.00
56	1	24.00	21.00	n.a.	n.a.	n.a.	n.a.	n.a.	n.a.
57	1	20.00	17.00	20.00	10.00	8.89	8.49	7.02	6.79
58	1	22.00	22.00	12.00	n.a.	n.a.	n.a.	n.a.	n.a.
59	1	20.00	22.00	15.38	2.00	4.00	6.00	3.00	3.00
60	1	21.00	23.00	11.00	9.00	10.00	8.00	7.00	4.00
61	1	17.00	17.00	15.00	n.a.	n.a.	n.a.	n.a.	n.a.
62	1	18.00	22.00	7.00	12.00	15.00	n.a.	n.a.	n.a.
63	1	23.00	26.00	24.00	n.a.	n.a.	n.a.	n.a.	n.a.

[a] Group codes: 0 = Placebo treatment, 1 = estrogen treatment

n.a. not available

8.4.3 More on "Beating the Blues": Checking the Model for the Correlation Structure

The model building carried out in the main text led us to fit a random intercept model to the BtB data. Histograms of the residuals suggested that the assumptions of normally distributed random terms were reasonable. Carry out further residual diagnostics to assess the linear mixed model assumptions of:

- Constant error variance across repeated measures
- Local independence

(Hint: Use the **Restructure Data Wizard** to convert the error residuals from long format into wide format and then assess the variances of and correlations between the residuals at the four time points. Note that under a model with effects for subjects, pairs of residuals from different time points are negatively correlated by definition and it is changes in the degree of these correlations that indicate a remaining correlation structure.)

Chapter 9

Logistic Regression: Who Survived the Sinking of the Titanic?

9.1 Description of Data

On April 14th, 1912, at 11.40 p.m., the Titanic, sailing from Southampton to New York, struck an iceberg and started to take on water. At 2.20 a.m. she sank; of the 2228 passengers and crew on board, only 705 survived. Data on Titanic passengers have been collected by many researchers, but here we shall examine part of a data set compiled by Thomas Carson. It is available on the Internet (http://hesweb1.med.virginia.edu/biostat/s/data/index.html). For 1309 passengers, these data record whether or not a particular passenger survived, along with the age, gender, ticket class, and the number of family members accompanying each passenger. Part of the SPSS Data View spreadsheet is shown in Display 9.1.

We shall investigate the data to try to determine which, if any, of the explanatory variables are predictive of survival. (The analysis presented in this chapter will be relatively straightforward; a far more comprehensive analysis of the Titanic survivor data is given in Harrell, 2001.)

9.2 Logistic Regression

The main question of interest about the Titanic survival data is: Which of the explanatory variables are predictive of the response, survived or died?

	pclass	survived	name	sex	age	sibsp	parch	ticket	fare	cabin	embarke
1	1	1	Allen, Miss. Elisabeth Walton	1	29	0	0	24160	211.3375	B5	1
2	1	1	Anderson, Mr. Harry	2	48	0	0	19952	26.5500	E12	1
3	1	0	Andrews, Mr. Thomas Jr	2	39	0	0	112050	.0000	A36	1
4	1	0	Artagaveytia, Mr. Ramon	2	71	0	0	PC 17609	49.5042		3
5	1	1	Aubart, Mme. Leontine Pauline	1	24	0	0	PC 17477	69.3000	B35	3
6	1	1	Barber, Miss. Ellen "Nellie"	1	26	0	0	19877	78.8500		1
7	1	1	Barkworth, Mr. Algernon Henry W	2	80	0	0	27042	30.0000	A23	1
8	1	0	Baumann, Mr. John D	2		0	0	PC 17318	25.9250		1
9	1	0	Baxter, Mr. Quigg Edmond	2	24	0	1	PC 17558	247.5208	B58 B60	3
10	1	1	Baxter, Mrs. James (Helene DeLa	1	50	0	1	PC 17558	247.5208	B58 B60	3
11	1	1	Bazzani, Miss. Albina	1	32	0	0	11813	76.2917	D15	3
12	1	0	Beattie, Mr. Thomson	2	36	0	0	13050	75.2417	C6	3
13	1	1	Behr, Mr. Karl Howell	2	26	0	0	111369	30.0000	C148	3
14	1	1	Bidois, Miss. Rosalie	1	42	0	0	PC 17757	227.5250		3
15	1	1	Bird, Miss. Ellen	1	29	0	0	PC 17483	221.7792	C97	1
16	1	0	Birnbaum, Mr. Jakob	2	25	0	0	13905	26.0000		3
17	1	1	Bissette, Miss. Amelia	1	35	0	0	PC 17760	135.6333	C99	1
18	1	1	Bjornstrom-Steffansson, Mr. Mau	2	28	0	0	110564	26.5500	C52	1
19	1	0	Blackwell, Mr. Stephen Weart	2	45	0	0	113784	35.5000	T	1
20	1	1	Blank, Mr. Henry	2	40	0	0	112277	31.0000	A31	3
21	1	1	Bonnell, Miss. Caroline	1	30	0	0	36928	164.8667	C7	1
22	1	1	Bonnell, Miss. Elizabeth	1	58	0	0	113783	26.5500	C103	1

Display 9.1   Characteristics of 21 Titanic passengers.

In essence, this is the same question that is addressed by the multiple regression model described in Chapter 4. Consequently, readers might ask: What is different here? Why not simply apply multiple regression to the Titanic data directly? There are two main reasons why this would not be appropriate:

■ The response variable in this case is binary rather than continuous. Assuming the usual multiple regression model for the probability of surviving could lead to predicted values of the probability outside the interval (0, 1).

■ The multiple regression model assumes that, given the values of the explanatory variables, the response variable has a normal distribution with constant variance. Clearly this assumption is not acceptable for a binary response.

Consequently, an alternative approach is needed and is provided by *logistic regression*, a brief account of the main points of which is given in Box 9.1.

Box 9.1 Logistic Regression

■ In multiple linear regression (see Chapter 4), the expected value of a response variable, y, is modeled as a linear function of the explanatory variables:

$$E(y) = \beta_0 + \beta_1 x_1 + \ldots + \beta_q x_q$$

■ For a binary response taking the values 0 and 1 (died and survived), the expected value is simply the probability, p, that the variable takes the value one, i.e., the probability of survival.
■ We could model p directly as a linear function of the explanatory variables and estimate the regression coefficients by least squares, but there are two problems with this direct approach:

1. The predicted value of p, $\hat{p}$, given by the fitted model should satisfy $0 \leq \hat{p} \leq 1$; unfortunately using a linear predictor does not ensure that this is so (see Exercise 9.4.2 for an illustration of this problem).

2. The observed values do not follow a normal distribution with mean p, but rather what is know as a *Bernoulli distribution* (see Everitt, 2002b).

■ A more suitable approach is to model p indirectly via what is known as the *logit transformation* of p, i.e., $\ln[p/(1 - p)]$.
■ This leads to the logistic regression model given by

$$\ln \frac{p}{1-p} = \beta_0 + \beta_1 x_1 + \ldots + \beta_q x_q$$

(The betas here are not, of course, the same as the betas in the first bulleted point in this box.)
■ In other words, the *log-odds* of survival is modeled as a linear function of the explanatory variables.
■ The parameters in the logistic regression model can be estimated by maximum likelihood (see Collett, 2003) for details.
■ The estimated regression coefficients in a logistic regression model give the estimated change in the log-odds corresponding to a unit change in the corresponding explanatory variable conditional on the other explanatory variables remaining constant. The parameters are usually exponentiated to give results in terms of odds.
■ In terms of p, the logistic regression model can be written as

$$p = \frac{\exp(\beta_0 + \beta_1 x_1 + \ldots + \beta_q x_q)}{1 + \exp(\beta_0 + \beta_1 x_1 + \ldots + \beta_q x_q)}$$

This function of the linear predictor is known as the *logistic function*.

■ Competing models in a logistic regression can be formally compared by a *likelihood ratio (LR) test*, a *score test* or by *Wald's test*. Details of these tests are given in Hosmer and Lemeshow (2000).

■ The three tests are asymptotically equivalent but differ in finite samples. The likelihood ratio test is generally considered the most reliable, and the Wald test the least (see Therneau and Grambsch, 2000, for reasons), although in many practical applications the tests will all lead to the same conclusion.

This is a convenient point to mention that logistic regression and the other modeling procedures used in earlier chapters, analysis of variance and multiple regression, can all be shown to be special cases of the *generalized linear model* formulation described in detail in McCullagh and Nelder (1989). This approach postulates a linear model for a suitable transformation of the expected value of a response variable and allows for a variety of different error distributions. The possible transformations are known as *link functions*. For multiple regression and analysis of variance, for example, the link function is simply the identity function, so the expected value is modeled directly, and the corresponding error distribution is normal. For logistic regression, the link is the logistic function and the appropriate error distribution is the binomial. Many other possibilities are opened up by the generalized linear model formulation — see McCullagh and Nelder (1989) for full details and Everitt (2002b) for a less technical account.

9.3 Analysis Using SPSS

Our analyses of the Titanic data in Table 9.1 will focus on establishing relationships between the binary passenger outcome survival (measured by the variable **survived** with "1" indicating survival and "0" death) and five passenger characteristics that might have affected the chances of survival, namely:

■ Passenger class (variable **pclass**, with "1" indicating a first class ticket holder, "2" second class, and "3" third class)
■ Passenger age (**age** recorded in years)
■ Passenger gender (**sex**, with females coded "1" and males coded "2")
■ Number of accompanying parents/children (**parch**)
■ Number of accompanying siblings/spouses (**sibsp**)

Our investigation of the determinants of passenger survival will proceed in three steps. First, we assess (unadjusted) relationships between survival and each potential predictor variable singly. Then, we adjust these relationships for potential confounding effects. Finally, we consider the possibility of interaction effects between some of the variables.

As always, it is sensible to begin by using simple descriptive tools to provide initial insights into the structure of the data. We can, for example, use the **Crosstabs...** command to look at associations between categorical explanatory variables and passenger survival (for more details on cross-tabulation, see Chapter 3). Cross-tabulations of survival by each of the categorical predictor variables are shown in Display 9.2. The results show that in our sample of 1309 passengers the survival proportions were:

- Clearly decreasing for lower ticket classes
- Considerably higher for females than males
- Highest for passengers with one sibling/spouse or three parents/children accompanying them

To examine the association between age and survival, we can look at a scatterplot of the two variables, although given the nature of the survival variable, the plot needs to be enhanced by including a Lowess curve (for details of Lowess curve construction see Chapter 2, Display 2.22). The Lowess fit will provide an informal representation of the change in proportion of "1s" (= survival proportion) with age. Without the Lowess curve, it is not easy to evaluate changes in density along the two horizontal lines corresponding to survival (code = 1) and nonsurvival (code = 0). The resulting graph is shown in Display 9.3. Survival chances in our sample are highest for infants and generally decrease with age although the decrease is not monotonic, rather there appears to be a local minimum at 20 years of age and a local maximum at 50 years.

Although the simple cross-tabulations and scatterplot are useful first steps, they may not tell the whole story about the data when confounding or interaction effects are present among the explanatory variables. Consequently, we now move on (again using simple descriptive tools) to assess the associations between pairs of predictor variables. Cross-tabulations and grouped box plots (not presented) show that in our passenger sample:

- Males were more likely to be holding a third-class ticket than females.
- Males had fewer parents/children or siblings/spouses with them than did females.
- The median age was decreasing with lower passenger classes.
- The median number of accompanying siblings/spouses generally decreased with age.
- The median number of accompanying children/parents generally increased with age.

a)

Survived?* Passenger class Crosstabulation

			Passenger class			Total
			first class	second class	third class	
Survived?	no	Count	123	158	528	809
		% within Passenger class	38.1%	57.0%	74.5%	61.8%
	yes	Count	200	119	181	500
		% within Passenger class	61.9%	43.0%	25.5%	38.2%
Total		Count	323	277	709	1309
		% within Passenger class	100.0%	100.0%	100.0%	100.0%

b)

Survived?* Gender Crosstabulation

			Gender		Total
			male	female	
Survived?	no	Count	682	127	809
		% within Gender	80.9%	27.3%	61.8%
	yes	Count	161	339	500
		% within Gender	19.1%	72.7%	38.2%
Total		Count	843	466	1309
		% within Gender	100.0%	100.0%	100.0%

c)

Survived?* Number of parents/children aboard Crosstabulation

			Number of parents/children aboard								Total
			0	1	2	3	4	5	6	9	
Survived?	no	Count	666	70	56	3	5	5	2	2	809
		% within Number of parents/children aboard	66.5%	41.2%	49.6%	37.5%	83.3%	83.3%	100.0%	100.0%	61.8%
	yes	Count	336	100	57	5	1	1			500
		% within Number of parents/children aboard	33.5%	58.8%	50.4%	62.5%	16.7%	16.7%			38.2%
Total		Count	1002	170	113	8	6	6	2	2	1309
		% within Number of parents/children aboard	100.0%	100.0%	100.0%	100.0%	100.0%	100.0%	100.0%	100.0%	100.0%

d)

Survived?* Number of siblings/spouses aboard Crosstabulation

			Number of siblings/spouses aboard							Total
			0	1	2	3	4	5	8	
Survived?	no	Count	582	156	23	14	19	6	9	809
		% within Number of siblings/spouses aboard	65.3%	48.9%	54.8%	70.0%	86.4%	100.0%	100.0%	61.8%
	yes	Count	309	163	19	6	3			500
		% within Number of siblings/spouses aboard	34.7%	51.1%	45.2%	30.0%	13.6%			38.2%
Total		Count	891	319	42	20	22	6	9	1309
		% within Number of siblings/spouses aboard	100.0%	100.0%	100.0%	100.0%	100.0%	100.0%	100.0%	100.0%

Display 9.2   Cross-tabulation of passenger survival by (a) passenger class, (b) gender, (c) number of parents/children aboard, and (d) number of siblings/spouses aboard.

As a result of these passenger demographics, confounding of variable effects is a very real possibility. To get a better picture of our data, a multi-way classification of passenger survival within strata defined by explanatory

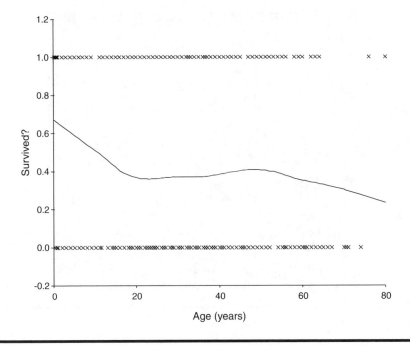

Display 9.3 Scattergraph of binary survival against age enhanced by inclusion of a Lowess curve.

variable-level combinations might be helpful. Before such a table can be constructed, the variables age, parch, and sibsp need to be categorized in some sensible way. Here we create two new variables, age_cat and marital, which categorize passengers into children (age <21 years) and adults (age ≥21 years) and into four marital status groups (1 = no siblings/spouses and no parents/children, 2 = siblings/spouses but no parents/children, 3 = no siblings/spouses but parents/children, and 4 = siblings/spouses and parents/children). The Recode command (see Chapter 3, Display 3.1) and the Compute command, in conjunction with the If Cases sub-dialogue box (see Chapter 1, Display 1.13) that allows sequential assignment of codes according to conditions, can be used to generate the new variables. The Crosstabs dialogue box is then employed to generate the required five-way table. (Note that the sequential Layer setting has to be used to instruct SPSS to construct cells for each level combination of the five variables, see Display 9.4.)

The resulting five-way tabulation is shown in Display 9.5. (We have edited the table somewhat to reduce its size. In SPSS, table editing is possible after double-clicking in the output table. Cells are missing in the table where no passenger survived for a particular combination of variables.) The stratified survival percentages are effectively a presentation of the adjusted associations between survival and potential predictors, adjusting

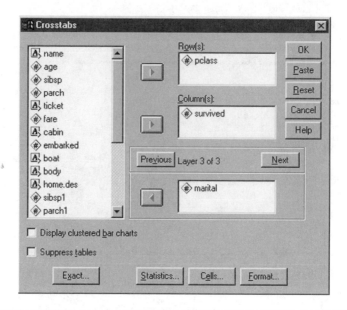

Display 9.4 Defining a five-way table.

each predictor for the effects of the other predictors included in the table. The stratified percentages also allow the informal assessment of interaction effects. Concentrating for the moment on the no siblings/spouses and no parents/children marital category to which most passengers (60%) belong, we find that in our sample:

■ Females with first-class tickets were most likely to survive.
■ Males and lower-class ticket holders were less likely to survive, although for females the survival chances gradually increased for higher-class tickets while for males only first-class tickets were associated with an increase.
■ Children were somewhat more likely to survive than adults.

We can now proceed to investigate the associations between survival and the five potential predictors using logistic regression. The SPSS logistic regression dialogue box is obtained by using the commands

Analyze – Regression – Binary Logistic...

We begin by looking at the unadjusted effects of each explanatory variable and so include a single explanatory variable in the model at a time. We start with the categorical explanatory variable pclass (Display 9.6):

Passenger class * Survived? * Gender * AGE_CAT * MARITAL Crosstabulation

% within Passenger class

MARITAL	AGE_CAT	Gender			Survived? yes
no sibs./spouse and no parents/childr.	Child	female	Passenger class	1	100.0%
				2	88.9%
				3	62.1%
		male	Passenger class	1	27.6%
				2	13.8%
				3	11.0%
	Adult	female	Passenger class	1	95.7%
				2	84.8%
				3	47.6%
		male	Passenger class	1	30.4%
				2	9.2%
				3	16.8%
sibs./spouse but no parents/childr.	Child	female	Passenger class	1	100.0%
				2	100.0%
				3	55.0%
		male	Passenger class	3	13.0%
	Adult	female	Passenger class	1	97.0%
				2	75.0%
				3	40.0%
		male	Passenger class	1	42.9%
				2	3.4%
				3	8.3%
no sibs./spouse but parents/childr.	Child	female	Passenger class	1	100.0%
				2	100.0%
				3	57.1%
		male	Passenger class	1	100.0%
				2	100.0%
				3	71.4%
	Adult	female	Passenger class	1	100.0%
				2	100.0%
				3	46.2%
		male	Passenger class	1	25.0%
				3	20.0%
sibs./spouses and parents/children	Child	female	Passenger class	1	66.7%
				2	90.9%
				3	34.2%
		male	Passenger class	1	75.0%
				2	88.9%
				3	19.2%
	Adult	female	Passenger class	1	95.2%
				2	93.8%
				3	37.5%
		male	Passenger class	1	33.3%
				2	9.1%
				3	10.0%

Display 9.5 Tabulation of survival by passenger class, gender, age, and marital status.

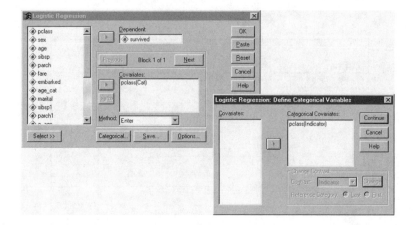

Display 9.6   Defining a logistic regression on one categorical explanatory variable.

- The binary dependent variable is declared under the **Dependent** list and the single explanatory variable under the **Covariates** list.
- By default, SPSS assumes explanatory variables are measured on an interval scale. To inform SPSS about the categorical nature of variable **pclass**, the **Categorical...** button is checked and **pclass** included in the **Categorical Covariates** list on the resulting **Define Categorical Variables** sub-dialogue box (Display 9.6). This will result in the generation of appropriate dummy variables. By default $k - 1$ indicator variables are generated for k categories with the largest category code representing the reference category (for more on dummy variable coding, see Chapter 4). It is possible to change the reference category or the contrast definition.
- We also check **CI for exp(B)** on the **Options** sub-dialogue box so as to include confidence intervals for the odds ratios in the output.

The resulting output is shown in Display 9.7 through Display 9.9. There are basically three parts to the output and we will discuss them in turn. The first three tables shown in Display 9.7 inform the user about the sample size, the coding of the dependent variable (SPSS will code the dependent variable 0-1 and model the probabilities of "1s" if the variable has different category codes), and the dummy variable coding of the categorical predictor variables. Here with only one categorical explanatory variable, we see that dummy variable (1) corresponds to first class, variable (2) to second class, and third class represents the reference category.

SPSS automatically begins by fitting a *null model,* i.e., one containing only an intercept parameter. Respective tables, shown in Display 9.8, are provided in the **Output Viewer** under the heading "Block 0: beginning block."

Case Processing Summary

Unweighted Cases[a]		N	Percent
Selected Cases	Included in Analysis	1309	100.0
	Missing Cases	0	.0
	Total	1309	100.0
Unselected Cases		0	.0
Total		1309	100.0

a. If weight is in effect, see classification table for the total number of cases.

Dependent Variable Encoding

Original Value	Internal Value
no	0
yes	1

Categorical Variables Codings

		Frequency	Parameter coding (1)	Parameter coding (2)
Passenger class	first class	323	1.000	.000
	second class	277	.000	1.000
	third class	709	.000	.000

Display 9.7 Variable coding summary output from logistic regression of survival on passenger class.

Classification Table [a,b]

			Predicted		
			Survived?		Percentage
Observed			no	yes	Correct
Step 0	Survived?	no	809	0	100.0
		yes	500	0	.0
	Overall Percentage				61.8

a. Constant is included in the model.
b. The cut value is .500

Variables in the Equation

		B	S.E.	Wald	df	Sig.	Exp(B)
Step 0	Constant	-.481	.057	71.548	1	.000	.618

Variables not in the Equation

			Score	df	Sig.
Step 0	Variables	PCLASS	127.856	2	.000
		PCLASS(1)	102.220	1	.000
		PCLASS(2)	3.376	1	.066
	Overall Statistics		127.856	2	.000

Display 9.8 Null model output from logistic regression of survival.

Classification Table[a]

			Predicted		
			Survived?		Percentage
	Observed		no	yes	Correct
Step 1	Survived?	no	686	123	84.8
		yes	300	200	40.0
	Overall Percentage				67.7

a. The cut value is .500

Omnibus Tests of Model Coefficients

		Chi-square	df	Sig.
Step 1	Step	127.765	2	.000
	Block	127.765	2	.000
	Model	127.765	2	.000

Variables in the Equation

		B	S.E.	Wald	df	Sig.	Exp(B)	95.0% C.I.for EXP(B)	
								Lower	Upper
Step 1[a]	PCLASS			120.536	2	.000			
	PCLASS(1)	1.557	.143	117.934	1	.000	4.743	3.581	6.282
	PCLASS(2)	.787	.149	27.970	1	.000	2.197	1.641	2.941
	Constant	-1.071	.086	154.497	1	.000	.343		

a. Variable(s) entered on step 1: PCLASS.

Display 9.9　Output from logistic regression of survival on passenger class.

The first two tables give details of the fit of this null model. In general these tables are of little interest, but it is helpful at this stage to describe just what they contain.

The first part of Display 9.8 is a "Classification Table" for the null model that compares survival predictions made on the basis of the fitted model with the true survival status of the passengers. On the basis of the fitted model, passengers are predicted to be in the survived category if their predicted survival probabilities are above 0.5 (different classification cut-off values can be specified in the **Options** sub-dialogue box). Here the overall survival proportion (0.382) is below the threshold and all passengers are classified as nonsurvivors by the null model leading to 61.8% (the nonsurvivors) being correctly classified.

Next, the "Variables in the Equation" table provides the Wald test for the null hypothesis of zero intercept (or equal survival and nonsurvival proportions); this test is rarely of any interest. Finally, the "Variables not in the Equation" table lists score tests for the variables not yet included in the model, here **pclass**. It is clear that survival is significantly related to passenger class (Score test: $X^2(2) = 127.9$, $p < 0.001$). Also shown are score tests for specifically comparing passenger classes with the reference category (third class).

The final parts of the SPSS output, presented in the **Output Viewer** under the heading "Block 1: Method = Enter" and given in Display 9.9 provide details of the requested logistic regression model. The latest "Classification Table" in Display 9.9 shows that inclusion of the **pclass** factor increases the percentage of correct classification by about 6 to 67.7%.

In addition to the tables of the previous output block, SPSS now displays a table labeled "Omnibus Tests of Model Coefficients." Under row heading "Block," this table contains the likelihood ratio (LR) test for comparing the model in the previous output block with the latest one, in other words, it is a test for assessing the effect of **pclass**. (Also given are an LR test for the whole model and an LR test for each term added in each step within block 1. Since our model consists of only one factor and the only step carried out is adding this factor, the three tests are identical.) We again detect a significant effect of passenger class (LR test: $X^2(2) = 127.8$, $p < 0.001$).

Finally, the latest "Variables in the Equation" table provides Wald's tests for all the variables included in the model. Consistent with the LR and score tests, the effect of **pclass** tests significant ($X^2(2) = 120.5$, $p < 0.001$). The parameter estimates (log-odds) are also given in the column labeled "B," with the column "S.E." providing the standard errors of these estimates. Since effects in terms of log-odds are hard to interpret, they are generally exponentiated to give odds-ratios (see columns headed "Exp(B)"). On our request 95% confidence intervals for the odds-ratios were added to the table. Comparing each ticket class with the third class, we estimate that the odds of survival were 4.7 times higher for first class passengers (CI form 3.6 to 6.3) and 2.2 times higher for second class passengers (1.6 to 2.9). Clearly, the chances of survival are significantly increased in the two higher ticket classes.

The results for the remaining categorical explanatory variables considered individually are summarized in Table 9.1. Since the observed survival frequencies of passengers accompanied by many family members were zero (see Display 9.2c and d), variables **sibsp** and **parch** were recoded before applying logistic regression. Categories 3 and higher were merged into single categories and survival percentages increased (new variables **sibsp1** and **parch1**). Table 9.1 shows that the largest increase in odds is found when comparing the two gender groups — the chance of survival among female passengers is estimated to be 8.4 times that of males.

It remains to look at the (unadjusted) effect of age. The shape of the Lowess curve plotted in Display 9.2 suggests that the survival probabilities might not be monotonically decreasing with age. Such a possibility can be modeled by using a third order polynomial for the age effect. To avoid multicollinearities, we center age by its mean (30 years) before calculating the linear (**c_age**), quadratic (**c_age2**), and cubic (**c_age3**) age terms (a similar approach was taken in the regression analysis of temperatures in Chapter 4). The three new age variables are then divided by their respective

Table 9.1 Unadjusted Effects of Categorical Predictor Variables on Survival Obtained from Logistic Regressions

Categorical Predictor	LR Test	OR for Survival	95% CI for OR
Passenger class (pclass)	$X^2(2) = 127.8,$		
First vs. third class	$p < 0.001$	4.743	3.581–6.282
Second vs. third class		2.197	1.641–2.941
Gender (sex)	$X^2(1) = 231.2,$		
Female vs. male	$p < 0.001$	8.396	6.278–11.229
Number of siblings/spouses	$X^2(3) = 14.2,$		
aboard (sibsp1)	$p < 0.001$		
0 vs. 3 or more		2.831	1.371–5.846
1 vs. 3 or more		5.571	2.645–11.735
2 vs. 3 or more		4.405	1.728–11.23
Number of parents/children	$X^2(3) = 46.7,$		
aboard (parch1)	$p < 0.001$		
0 vs. 3 or more		1.225	0.503–2.982
1 vs. 3 or more		3.467	1.366–8.802
2 vs. 3 or more		2.471	0.951–6.415

standard deviations (14.41, 302.87, and 11565.19) simply to avoid very small regression coefficients due to very large variable values. Inclusion of all three age terms under the **Covariates** list in the **Logistic Regression** dialogue box gives the results shown in Display 9.10. We find that the combined age terms have a significant effect on survival (LR: $X^2(3) = 16.2$, $p = 0.001$). The single parameter Wald tests show that the quadratic and cubic age terms contribute significantly to explaining variability in survival probabilities. These results confirm that a linear effect of age on the log-odds scale would have been too simplistic a model for these data.

Having found that all our potential predictors are associated with survival when considered singly, the next step is to model their effects simultaneously. In this way, we will be able to estimate the effect of each adjusted for the remainder. A logistic regression model that accounts for the effects of all five explanatory variables can be fitted by including the four categorical variables and the three age terms under the **Covariates** list of the **Logistic Regression** dialogue box (see Display 9.6). (The **Categorical...** button was again employed to define categorical variables.) Part of the resulting output is shown in Display 9.11. We note from the "Case Processing Summary" table that the number of cases included in the analysis is reduced to 1046 because information on at least one explanatory variable (age) is missing for 263 passengers.

The SPSS output does not supply an LR test for the effect of each variable when added to a model containing all the other explanatory variables, rather the table of Omnibus Tests provides a test for the effect

Omnibus Tests of Model Coefficients

		Chi-square	df	Sig.
Step 1	Step	16.153	3	.001
	Block	16.153	3	.001
	Model	16.153	3	.001

Variables in the Equation

		B	S.E.	Wald	df	Sig.	Exp(B)	95.0% C.I.for EXP(B) Lower	Upper
Step 1[a]	C_AGE	.098	.113	.742	1	.389	1.102	.883	1.376
	C_AGE2	.235	.074	10.155	1	.001	1.265	1.095	1.461
	C_AGE3	-.339	.128	6.999	1	.008	.713	.555	.916
	Constant	-.501	.079	40.372	1	.000	.606		

a. Variable(s) entered on step 1: C_AGE, C_AGE2, C_AGE3.

Display 9.10   LR test and odds ratios for logistic regression of survival on three age terms.

Case Processing Summary

Unweighted Cases[a]		N	Percent
Selected Cases	Included in Analysis	1046	79.9
	Missing Cases	263	20.1
	Total	1309	100.0
Unselected Cases		0	.0
Total		1309	100.0

a. If weight is in effect, see classification table for the total number of cases.

Variables in the Equation

		B	S.E.	Wald	df	Sig.	Exp(B)	95.0% C.I.for EXP(B) Lower	Upper
Step 1[a]	PCLASS			85.200	2	.000			
	PCLASS(1)	2.159	.234	85.195	1	.000	8.661	5.476	13.699
	PCLASS(2)	.874	.206	17.950	1	.000	2.397	1.600	3.591
	SEX(1)	2.550	.176	210.422	1	.000	12.812	9.077	18.083
	C_AGE	-.413	.156	6.955	1	.008	.662	.487	.899
	C_AGE2	.195	.106	3.363	1	.067	1.215	.987	1.497
	C_AGE3	-.200	.165	1.474	1	.225	.819	.592	1.131
	SIBSP1			18.725	3	.000			
	SIBSP1(1)	2.166	.510	18.031	1	.000	8.722	3.210	23.700
	SIBSP1(2)	2.008	.516	15.152	1	.000	7.449	2.710	20.475
	SIBSP1(3)	1.541	.653	5.562	1	.018	4.668	1.297	16.798
	PARCH1			5.525	3	.137			
	PARCH1(1)	.518	.552	.880	1	.348	1.679	.569	4.955
	PARCH1(2)	1.031	.583	3.126	1	.077	2.805	.894	8.801
	PARCH1(3)	.844	.623	1.838	1	.175	2.326	.686	7.884
	Constant	-4.989	.755	43.661	1	.000	.007		

a. Variable(s) entered on step 1: PCLASS, SEX, C_AGE, C_AGE2, C_AGE3, SIBSP1, PARCH1.

Display 9.11   Selected output from logistic regression of survival on all potential explanatory variables.

of all explanatory variables simultaneously. The relevant LR tests can be generated by careful stepwise model building using the blocking facility (see later), but here we resort to Wald tests that are generated by default (Display 9.11). The final "Variables in the Equation" table shows that the variable parch1 does not contribute significantly to explaining survival probabilities once the other variables are accounted for (Wald: $X^2(3) = 5.5$, $p = 0.14$) and so we refit the model excluding this variable. The new model output (not shown) indicates that the third order age term is no longer necessary (Wald: $X^2(1) = 2.6$, $p = 0.11$) and so this term is also dropped.

The final main effects model (not shown) contains terms for age, passenger class, gender, and number of siblings/spouses, each of which contributes significantly at the 5% level after adjusting for other model terms. We will leave interpretation of the model until we have investigated the possibility of interaction effects, but it is worth noting the changes in odds ratios compared with the unadjusted estimates in Table 9.1 and Display 9.10 as a result of adjusting for potential confounding effects.

There are a number of modeling strategies that might be applied to selecting a final model for our data (see, for example, Kleinbaum and Klein, 2002). We opt for the simple approach of only considering two-way interactions of model terms that had each been demonstrated to have a significant main effect on survival. A potential pitfall of this approach is that higher order interaction effects that do not coincide with lower order effects might be missed (a more comprehensive model building approach can be found in Harrell, 2001).

We will test the two-way interaction terms one-by-one using the blocking facility to add the interaction term of interest to the main effects model identified previously. Display 9.12 illustrates this for the age by sex

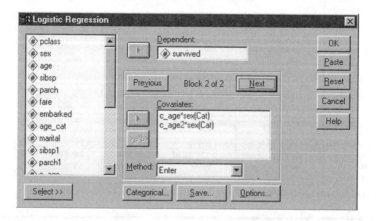

Display 9.12   Adding interaction effects to a logistic regression model.

Omnibus Tests of Model Coefficients

		Chi-square	df	Sig.
Step 1	Step	23.869	2	.000
	Block	23.869	2	.000
	Model	477.239	10	.000

Display 9.13 LR test output for age by gender interaction effect on survival probabilities.

interaction (for more on blocking see Chapter 4, Display 4.15). An interaction effect can be defined in the **Logistic Regression** dialogue box by highlighting the variables involved and clicking the **>a*b>** symbol. This automatically creates the relevant interaction terms treating contributing variables as continuous or categorical depending on how the main effects (which have to be in the model) are defined.

As a consequence of the blocking, the "Omnibus Tests of Model Coefficients" table shown in the **Output Viewer** under the heading "Block 2: Method = Enter" now contains an LR test that compares the model including the interaction effect in question with a model containing only main effects. The LR test for assessing the gender by age interaction terms is shown in Display 9.13 under the row titled "Block." The table indicates that two parameters have been added. The first allows the effect of the linear age term to vary with gender, the second does the same for the quadratic age term. It appears that an age by gender interaction *is* helpful in explaining survival probabilities. We proceed to test the remaining two-way interactions in the same way, each time including only the interaction term in question in the second block.

During this process, we came across an apparent problem when including the **sibsp1×pclass** interaction, namely, odds ratio estimates of less than 0.01 with accompanying confidence intervals ranging from zero to infinity. This is not an uncommon problem in logistic regression when observed proportions in factor level combinations are extreme (near zero or one). In such circumstances, the algorithm for finding the maximum likelihood solution often suffers convergence problems. A cross-tabulation of **survived** and **sibsp1** categories within layers defined by **pclass** (not shown) confirms that this is indeed the case here. In the second passenger class, every passenger with three or more spouses/siblings survived. Out of necessity, we therefore further merge the **sibsp1** variable into three categories, combining the two higher categories into "2 or more" (new variable **sibsp2**) and then use this variable in the logistic modeling.

The results of the interaction testing are summarized in Table 9.2, indicating that interactions between gender and passenger class, gender and age, passenger class and number of siblings/spouses, and age and

Table 9.2   LR Test Results for Assessing Two-Way Interaction Effects on Survival Probabilities

Interaction Involving		LR Test		
Variable 1	Variable 2	Deviance Change	DF	p-Value
pclass	sex	52.8	2	<0.001
c_age, c_age2	sex	23.9	2	<0.001
sibsp2	sex	0.84	2	0.66
c_age, c_age2	pclass	7.4	4	0.12
sibsp2	pclass	16.6	4	0.002
c_age, c_age2	sibsp2	9.7	4	0.045

Variables in the Equation

		B	S.E.	Wald	df	Sig.	Exp(B)	95.0% C.I.for EXP(B)	
								Lower	Upper
Step 1[a]	PCLASS			14.394	2	.001			
	PCLASS(1)	3.974	1.102	13.003	1	.000	53.214	6.136	461.499
	PCLASS(2)	1.736	.950	3.338	1	.068	5.673	.881	36.521
	SEX(1)	1.831	.280	42.854	1	.000	6.238	3.606	10.793
	C_AGE	-.870	.128	45.865	1	.000	.419	.326	.539
	C_AGE2	.343	.105	10.769	1	.001	1.409	1.148	1.730
	SIBSP2			19.103	2	.000			
	SIBSP2(1)	1.992	.456	19.102	1	.000	7.333	3.001	17.919
	SIBSP2(2)	1.734	.485	12.782	1	.000	5.663	2.189	14.651
	PCLASS * SEX			28.309	2	.000			
	PCLASS(1) by SEX(1)	1.911	.615	9.644	1	.002	6.757	2.023	22.567
	PCLASS(2) by SEX(1)	2.390	.478	24.957	1	.000	10.915	4.273	27.878
	C_AGE by SEX(1)	.423	.214	3.900	1	.048	1.526	1.003	2.322
	C_AGE2 by SEX(1)	-.233	.199	1.369	1	.242	.792	.536	1.171
	PCLASS * SIBSP2			8.409	4	.078			
	PCLASS(1) by SIBSP2(1)	-2.500	1.110	5.070	1	.024	.082	.009	.723
	PCLASS(1) by SIBSP2(2)	-1.975	1.138	3.011	1	.083	.139	.015	1.291
	PCLASS(2) by SIBSP2(1)	-1.953	.975	4.013	1	.045	.142	.021	.959
	PCLASS(2) by SIBSP2(2)	-1.758	1.027	2.928	1	.087	.172	.023	1.291
	Constant	-3.920	.479	67.024	1	.000	.020		

a. Variable(s) entered on step 1: PCLASS * SEX , C_AGE * SEX , C_AGE2 * SEX , PCLASS * SIBSP2 .

Display 9.14   Parameter estimates from final model for survival probabilities.

number of siblings/spouses should be considered for inclusion in our logistic regression model. Fitting such a model followed by an examination of the relevant Wald tests demonstrates that all interaction terms except the age by number of siblings/spouses were significant at the 10% level after adjusting for other interaction terms. So in our chosen final model shown in Display 9.14, this particular interaction term is excluded.

It is quite an elaborate process to construct odds ratios for the effect of predictor variables when the variables are involved in interactions. For example, the "Exp(B)" value for PCLASS(1) in Display 9.14, 53.2 represents the odds ratio of survival comparing first-class with third-class ticket holders for passengers falling into the reference categories of all the factors interacting with **pclass** (i.e., for passengers of the second **sex** category "male" and the third **sibsp2** category "2 or more siblings/spouses"). Careful recoding and changing of the reference categories is necessary to generate a complete list of comparisons (for more details on parameterizations in the presence of interactions, see Kleinbaum and Klein, 2002).

As an alternative means for interpreting our fitted logistic model, we construct a graphical display of the log-odds of survival. We opt for displaying predicted survival probabilities on the log-odds scale since the logistic model assumes additive effects of the explanatory variables on this scale. The scatterplots shown in Display 9.15 can be generated using the following instructions:

- Save passengers' predicted survival probabilities as a new variable, **pre_1**, in the **Data View** spreadsheet by ticking **Predicted Values: Probabilities** in the **Save New Variables** sub-dialogue box when fitting the final logistic regression model.
- Transform these values into odds using the formula **odds = pre_1/(1 − pre_1)** and calculate a log-odds variable, **ln_odds** by using the formula **ln_odds = ln(odds)**.
- Generate a passenger class and gender combination factor (**class.se**) by using the **Compute** command with the **Numeric Expression 100 × pclass + 1 × sex** (see Chapter 1, Display 1.13). This results in a factor with six levels. Each level code is a three-digit number; with the first digit indicating the passenger class, the last the gender, and middle digit zero.
- Use the **Split File** command to organize output by groups defined by **sibsp2** (see Chapter 1, Display 1.10).
- Use the **Simple Scatterplot** dialogue box to produce a scatterplot of **ln_odds** against **age** with markers defined by **class.se** (see Chapter 2, Display 2.20).
- Open the **Chart Editor** and employ the commands **Chart – Options – Fit Line: Subgroups – Quadratic regression** from the menu bar to fit quadratic polynomials to the predicted values within each factor level combination.
- Carry out further editing of the chart to improve its appearance.

Display 9.15 illustrates the detected relationships between passenger characteristics and survival of the sinking of the Titanic:

- *Identified predictors*: Each of the variables, passenger age, gender, class of ticket, and number of accompanying siblings/spouses, make an independent contribution to predicting the chance of

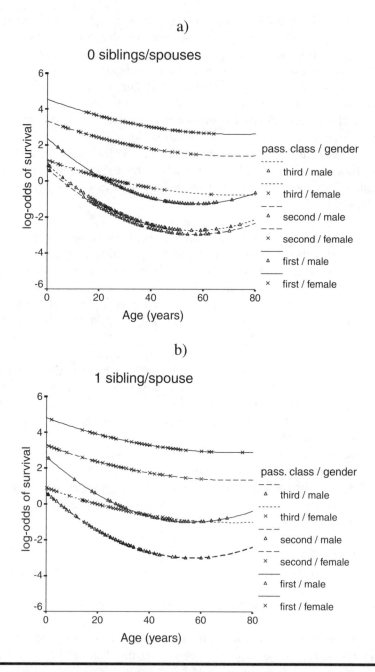

Display 9.15 **Log-odds of survival predicted by the final logistic model by passenger age, gender, ticket class, and number of accompanying siblings/spouses. The symbols indicate predictions for passenger characteristics observed in the sample, the curves the extrapolated values.**

c)

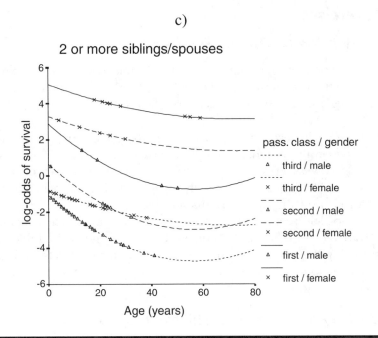

Display 9.15 (continued).

survival. Young (<20 years), female, first-class ticket holders are predicted to have the highest chance of survival and male third-class ticket holders in their twenties and thirties accompanied by two or more siblings/spouses the lowest.

■ *Age by gender interaction*: Survival chances are predicted to generally decrease with increasing passenger age and to be lower for males. However, the rate of decrease depends on gender with the survival chances of male children decreasing more rapidly with every additional year of age. Thus, the survival advantage of females over males appears to widen with increasing age.

■ *Gender by ticket class interaction*: Survival chances of females are predicted to be increasing gradually with higher passenger class, while for the majority of males (those accompanied by none or one sibling/spouse), only those carrying first-class tickets are predicted to have an increased chance of survival.

■ *Tentative ticket class by number of siblings/spouses interaction*: The predicted difference in chance of survival between holders of different classes of tickets is largest for passengers accompanied by two or more siblings/spouses with third-class ticket holders, with larger families predicted to have particularly low chances of survival.

9.4 Exercises

9.4.1 More on the Titanic Survivor Data

The presentation of the analyses of the Titanic data in the main text summarized the results from a number of methods. Generate the full SPSS output underlying these summaries; specifically:

- Use simple descriptive tools to assess the possibility of confounding between the potential predictor variables.
- Use Lowess curves to visualize the potential effect of age on survival and interactions involving age, gender, and passenger class for each of the marital status groups.
- Fit logistic regression models of survival with single explanatory variables to confirm the unadjusted effects shown in Table 9.1.
- Fit logistic regression models including main effects and single interaction terms to confirm the test results shown in Table 9.2.

9.4.2 GHQ Scores and Psychiatric Diagnosis

The data shown below were collected during a study of a psychiatric screening questionnaire, the GHQ (Goldberg, 1972), designed to help identify possible psychiatric cases. Use these data to illustrate why logistic regression is superior to linear regression for a binary response variable.

(Hint: Use the **Weight Cases** command to inform SPSS about the cross-classified format of the data.)

9.4.3 Death Penalty Verdicts Revisited

The data shown in Table 3.5 on race and the death penalty in the U.S. were analyzed in Chapter 3, Section 3.3.7 by means of the Mantel-Haenszel test. This test procedure allowed us to assess the association between the use of the death penalty in a murder verdict and the race of the defendant within strata defined by the victim's race. Use logistic regression of the binary outcome "death penalty" to estimate the (unadjusted) ORs for defendant's race and victim's race and the ORs of each of the factors after adjusting for the other. Compare OR estimates and confidence intervals with those constructed in Section 3.3.7.

Table 9.3   Caseness by GHQ Score and Gender (*n* = 278)

		Caseness Distribution	
GHQ Score	Sex (0 = Female, 1 = Male)	Number of Cases	Number of Non Cases
0	0	4	80
1	0	4	29
2	0	8	15
3	0	6	3
4	0	4	2
5	0	6	1
6	0	3	1
7	0	2	0
8	0	3	0
9	0	2	0
10	0	1	0
0	1	1	36
1	1	2	25
2	1	2	8
3	1	1	4
4	1	3	1
5	1	3	1
6	1	2	1
7	1	4	2
8	1	3	1
9	1	2	0
10	1	2	0

Chapter 10

Survival Analysis: Sexual Milestones in Women and Field Dependency of Children

10.1 Description of Data

A subset of data recorded in a study of the relationship between pubertal development, sexual milestones, and childhood sexual abuse in women with eating disorders (Schmidt et al., 1995) is shown in Table 10.1. The data that will be considered in this chapter are times to first sexual intercourse for two groups of women: female out-patients (Cases) of an Eating Disorder Unit who fulfilled the criteria for Restricting Anorexia Nervosa (RAN), and female polytechnic students (Controls). Some women had not experienced sex by the time they filled in the study questionnaire, so their event times are *censored* at their age at the time of the study (see Box 10.1). Such observations are marked by a "+" in Table 10.1. For these data the main interest is in testing whether the time to first sexual intercourse is the same in both groups.

The second set of data shown in Table 10.2 was collected from a sample of primary school children. Each child completed the Embedded Figures Test (EFT; Witkin, Oltman, Raskin, and Karp, 1971), which measures

Table 10.1   Times to First Sexual Intercourse

Subject	Group (0 = control, 1 = case)	Time (years)	Age at the Time of the Study (years)	Status
1	1	30+	30	Censored
2	1	24+	24	Censored
3	1	12	18	Event
4	1	21	42	Event
5	1	19+	19	Censored
6	1	18	39	Event
7	1	24	30	Event
8	1	20	30	Event
9	1	24	33	Event
10	1	28	38	Event
11	1	17+	17	Censored
12	1	18+	18	Censored
13	1	18+	18	Censored
14	1	27+	27	Censored
15	1	31+	31	Censored
16	1	17+	17	Censored
17	1	28+	28	Censored
18	1	29+	29	Censored
19	1	23	23	Event
20	1	19	35	Event
21	1	19	28	Event
22	1	18+	18	Censored
23	1	26+	26	Censored
24	1	22+	22	Censored
25	1	20+	20	Censored
26	1	28+	28	Censored
27	1	20	26	Event
28	1	21+	21	Censored
29	1	18	22	Event
30	1	18+	18	Censored
31	1	20	25	Event
32	1	21+	21	Censored
33	1	17+	17	Censored
34	1	21+	21	Censored
35	1	16	22	Event
36	1	16	20	Event
37	1	18	21	Event
38	1	21	29	Event
39	1	17	20	Event
40	1	17	20	Event

Table 10.1 (continued)   Times to First Sexual Intercourse

Subject	Group (0 = control, 1 = case)	Time (years)	Age at the Time of the Study (years)	Status
41	0	15	20	Event
42	0	13	20	Event
43	0	15	20	Event
44	0	18	20	Event
45	0	16	19	Event
46	0	19	20	Event
47	0	14	20	Event
48	0	16	20	Event
49	0	17	20	Event
50	0	16	21	Event
51	0	17	20	Event
52	0	18	22	Event
53	0	16	22	Event
54	0	16	20	Event
55	0	16	38	Event
56	0	17	21	Event
57	0	16	21	Event
58	0	19	22	Event
59	0	19	36	Event
60	0	17	24	Event
61	0	18	30	Event
62	0	20	39	Event
63	0	16	20	Event
64	0	16	19	Event
65	0	17	22	Event
66	0	17	22	Event
67	0	17	23	Event
68	0	18+	18	Censored
69	0	16	29	Event
70	0	16	19	Event
71	0	19	22	Event
72	0	19	22	Event
73	0	18	21	Event
74	0	17	19	Event
75	0	19	21	Event
76	0	16	20	Event
77	0	16	22	Event
78	0	15	18	Event
79	0	19	26	Event
80	0	20	23	Event

Table 10.1 (continued)   Times to First Sexual Intercourse

Subject	Group (0 = control, 1 = case)	Time (years)	Age at the Time of the Study (years)	Status
81	0	16	20	Event
82	0	15	21	Event
83	0	17	21	Event
84	0	18	21	Event

field dependence, i.e., the extent to which a person can abstract the logical structure of a problem from its context (higher scores indicate less abstraction). Then, the children were randomly allocated to one of two experimental groups and timed as they constructed a 3 × 3 pattern from nine colored blocks, taken from the Wechsler Intelligence Scale for Children (WISC, Wechsler, 1974). The two groups differed in the instructions they were given for the task. The row group was told to start with a row of three blocks and the corner group was told to start with a corner of three blocks. The age of each child in months was also recorded. Children were allowed up to ten minutes to complete the task, after which time the experimenter intervened to show children who had not finished the correct solution. Such censored observations are shown as 600+ in the table. For this study, it is the effects of the covariates, EFT, age, and experimental group on time to completion of the task that are of interest.

10.2 Survival Analysis and Cox's Regression

Data sets containing censored observations such as those described in the previous section are generally referred to by the generic term *survival data* even when the endpoint of interest is not death but something else (as it is in both Tables 10.1 and 10.2). Such data require special techniques for analysis for two main reasons:

- Survival data are usually not symmetrically distributed. They will often be positively skewed, with a few people surviving a very long time compared with the majority, so assuming a normal distribution will not be reasonable.
- Censoring: At the completion of the study; some patients may not have reached the endpoint of interest. Consequently, the exact survival times are not known. All that is known is that the survival times are greater than some value and the observations are said to be censored (such censoring is more properly known as *right censoring*).

Table 10.2   Times to Completion of the WISC Task

Subject	Time (sec)	EFT Score	Age (months)	Group (0 = "row group", 1 = "corner group")	Status
1	600+	47.91	103.9	0	Censored
2	600+	54.45	77.3	0	Censored
3	600+	47.04	110.6	0	Censored
4	600+	50.07	106.9	0	Censored
5	600+	43.65	95.4	0	Censored
6	389.33	31.91	110.6	0	Completed
7	600+	55.18	101.1	0	Censored
8	290.08	29.33	81.7	0	Completed
9	462.14	23.15	110.3	0	Completed
10	403.61	43.85	86.9	0	Completed
11	258.7	26.52	97.5	0	Completed
12	600+	46.97	104.3	0	Censored
13	559.25	35.2	88.8	0	Completed
14	600+	45.86	96.9	0	Censored
15	600+	36.8	105.7	0	Censored
16	600+	58.32	97.4	0	Censored
17	600+	45.4	102.4	0	Censored
18	385.31	37.08	100.3	0	Completed
19	530.18	38.79	107.1	0	Completed
20	600+	63.2	91.1	0	Censored
21	490.61	38.18	79.6	0	Completed
22	473.15	29.78	101.7	0	Completed
23	545.72	39.21	106.3	0	Completed
24	600+	42.56	92.3	0	Censored
25	600+	55.53	112.6	0	Censored
26	121.75	22.08	90.3	1	Completed
27	589.25	50.16	115.5	1	Completed
28	472.21	56.93	117.3	1	Completed
29	600+	36.09	97	1	Censored
30	359.79	49.65	98.3	1	Completed
31	371.95	34.49	109.3	1	Completed
32	214.88	14.67	96.7	1	Completed
33	272.82	39.24	93.7	1	Completed
34	600+	32.19	122.9	1	Censored
35	563.55	34.16	98.1	1	Completed
36	600+	49.12	121.2	1	Censored
37	461.61	41.23	111.4	1	Completed
38	600+	57.49	101.3	1	Censored
39	600+	47.86	103.4	1	Censored
40	147.43	34.24	94	1	Completed

Table 10.2 (continued)   Times to Completion of the WISC Task

Subject	Time (sec)	EFT Score	Age (months)	Group (0 = "row group", 1 = "corner group")	Status
41	600+	47.36	93.1	1	Censored
42	600+	39.06	96	1	Censored
43	600+	55.12	109.7	1	Censored
44	429.58	47.26	96.5	1	Completed
45	223.78	34.05	97.5	1	Completed
46	339.61	29.06	99.6	1	Completed
47	389.14	24.78	120.9	1	Completed
48	600+	49.79	95.6	1	Censored
49	600+	43.32	81.5	1	Censored
50	396.13	56.72	114.4	1	Completed

Of central importance in the analysis of survival time data are two functions used to describe their distribution, namely, the *survival function* (*survivor function*) and the *hazard function*. Both are described in Box 10.1. To assess the effects of explanatory variables on survival times, a method known as *Cox's regression* is generally employed. This technique is described briefly in Box 10.2.

Box 10.1 Survival Function and Hazard Function

■ The survivor function, $S(t)$ is defined as the probability that the survival time T is greater than t, i.e.,

$$S(t) = \Pr(T > t)$$

■ When there is no censoring, the survivor function can be estimated as

$$\hat{S}(t) = \frac{\text{Number of individuals with survival times} > t}{\text{Number of individuals in the dataset}}$$

■ When there is censoring, the *Kaplan-Meier estimate* is generally used; this is based on the use of a series of conditional probabilities (see Everitt, 2002b) for details and a numerical example).

- To compare the survivor functions between groups, the *log-rank test* can be used. This test involves the calculation of the expected number of "deaths" for each failure time in the data set, assuming that the chances of dying, given that the subjects are at risk, are the same for all groups. The total number of expected deaths is then computed for each group by adding the expected number of deaths for each failure time. The log-rank test then compares the observed number of deaths in each group with the expected number of deaths using a chi-square test (see Hosmer and Lemeshow, 1999, for further details).
- In the analysis of survival data, it is often of interest to assess which periods have high or low classes of death (or whatever the event of interest might be) among those still at risk at the time.
- A suitable approach to characterize this is the hazard function, $h(t)$, defined as the probability that an individual experiences the event of interest in a small time interval s, given that the individuals have survived up to the beginning of the interval.
- The conditioning feature of this definition is very important. For example, the probability of dying at age 100 is very small because most people die before that age; in contrast, the probability of a person dying at age 100 who has reached that age is much greater.
- The hazard function is a measure of how likely an individual is to experience an event as a function of the age of the individual; it is often also known as the *instantaneous death rate*.
- Although it is easy to calculate the hazard function from sample data (see Everitt and Rabe-Hesketh, 2001). it is rarely done since the estimate is generally too noisy. The cumulative hazard function defined as the integral of the hazard function up to some time t is generally a more useful characterization of the instantaneous death rate aspect of survival time data.

Box 10.2 Cox's Regression

- Of most interest in many studies in which survival times are collected is to assess the effects of explanatory variables on these times.
- The main vehicle for modeling in this case is the hazard function.

■ But modeling the hazard function directly as a linear function of explanatory variables is not appropriate since $h(t)$ is restricted to being positive.

■ A more suitable model might be

$$\ln\left[h(t)\right] = \beta_0 + \beta_1 x_1 \ldots + \beta_q x_q$$

■ The problem with this is that it is only suitable for a hazard function that is constant over time — such a model is very restrictive since hazards that increase or decrease with time, or have some more complex form, are far more likely to occur in practice. In general, it may be difficult to find the appropriate explicit function of time to include.

■ The problem is overcome in the proportional hazards model proposed by Cox (1972), namely,

$$\ln\left[h(t)\right] = \ln\left[h_0(t)\right] + \beta_1 x_1 \ldots + \beta_q x_q$$

where $h_0(t)$ is known as the *baseline hazard function*, being the hazard function for individuals with all explanatory variables equal to zero. This function is left unspecified. (The estimated cumulative baseline hazard can be estimated from sample data and is often useful — see the text for more details.)

■ This model forces the hazard ratio between two individuals to be constant over time (see Everitt, 2002b). In other words, if an individual has a risk of death at some initial time point that is twice as high as another individual, then at all later times the risk of death remains twice as high.

■ Proportionality of hazards is an assumption that may not necessarily hold (see text for possible ways to detect departures from the assumption and alternatives when it is not thought to hold).

■ The model also assumes that the explanatory variables affect the hazard rate multiplicatively, i.e., the log hazard additively (see text for how the assumption can be assessed).

■ Parameters in Cox's regression are estimated by a form of maximum likelihood (for details, see Collett, 2003).

■ The interpretation of the parameter β_j is that $\exp(\beta_j)$ gives the hazard ratio, the factor change associated with an increase of one unit in x_j, all other explanatory variables remaining constant.

10.3 Analysis Using SPSS

10.3.1 Sexual Milestone Times

In the sexual milestone times study, the event of interest is the time of first sexual intercourse and so the survival times are the ages at first sex. For some women these times are censored since, at the time of the study, they had not experienced intercourse; all that is then known is that the time of first sex is greater than their age at the time of the study.

The format of the **Data View** spreadsheet for the times to first sexual intercourse is very similar to that of Table 10.1. The rows of the spreadsheet refer to women and the columns provide information on diagnostic group (factor **diagnos**), the women's age at first sex (**agesex**) (possibly censored), the women's age at the time of the study (**age**), and an event identifier (**sexstat**). The latter takes the value "1" when the event of interest, here first sex, has occurred by the time of the study (survival time not censored). Otherwise the value is "0," indicating censoring so that **age** and **agesex** will have the same value and we know only that age at first sex is greater than this value.

For this study, the main interest lies in comparing ages at first sex between the cases and the controls. Before undertaking a formal test, it might be useful to display the survival times graphically. One possibility is to display individual survival times and indicate the censoring in the plot in some way. Such a plot can be created in SPSS by constructing a *drop line chart*. The necessary instructions are as follows:

- Rearrange the spreadsheet rows in ascending order of sexstat and agesex by using the commands **Data – Sort Cases – Sort by: sexstat (A) agesex (A) – Sort Order Ascending – OK**.
- Use the **Compute** command (see Chapter 1, Display 1.13) to generate a new variable (labeled **start**) that contains zero values.
- Employ the **Split File** command to organize all further output by diagnostic groups (see Chapter 1, Display 1.10).
- The commands **Graphs – Line... – Drop-line – Values of individual cases – Define** open the **Define Drop-line: Values of Individual Cases** dialogue box where the variables **agesex**, and **start** need to be listed under the **Points Represent** list and the variable **sexstat** as the **Category Labels Variable**.

The resulting charts are shown in Display 10.1. It is apparent that almost all censoring occurs among the cases. Although ages at first sex appear younger in the control group based on the uncensored data, such comparisons can be misleading when differential censoring is present. It would help if we could display the distributions of the survival times in

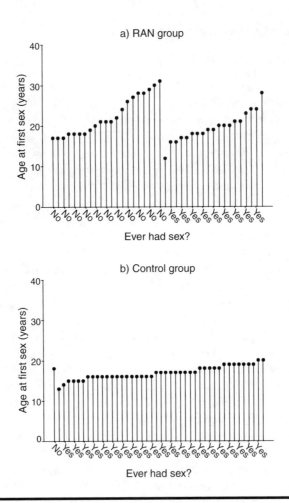

Display 10.1   Drop line charts of times to first sex.

the two groups satisfactorily, but standard summary displays for continuous data, such as box plots or histograms (see Chapter 2), are *not* suitable for this purpose since they do not take account of the censoring in the survival times. We need to describe the distributions in a way that allows for this censoring and this is just what the Kaplan-Meier estimate of the survival function allows.

Kaplan-Meier estimates of the survivor functions can be generated using the commands

Analyze – Survival – Kaplan-Meier...

This opens the **Kaplan-Meier** dialogue box where the (potentially right censored) survival time variable (**agesex**) is specified under the **Time** list

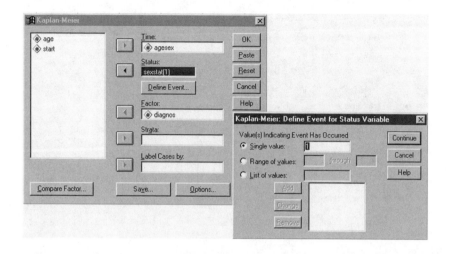

Display 10.2   **Generating Kaplan-Meier estimates of group-wise survival functions.**

and the censoring variable (**sexstat**) under the **Status** list (Display 10.2). SPSS allows for any variable to supply the censoring information and, therefore, requires use of the **Define Event...** button to specify which values of the censoring variable (here the value "1") indicate that the event has occurred. Kaplan-Meier curves can be constructed within categories of a variable listed under the **Factor** list. Here we request separate curves to be constructed for each group (factor **diagnos**).

The resulting Kaplan-Meier estimates are shown in Display 10.3. For each group, the observed times (values of **agesex**) are listed in ascending order under the column "Time" with the "Status" column noting whether or not the time value indicates the event of interest (first sex). For each observed event time, the Kaplan-Meier estimate of the probability of surviving longer than this time, or of first experiencing sex at a later age than that observed, is displayed (column "Cumulative Survival"). The standard errors of the Kaplan-Meier estimates are also provided (column "Standard Error"). Additional information given is the number of events that have occurred up to and including the current time (column "Cumu-lative events") and the remaining number of subjects in the group with time observations (column "Number Remaining"). The latter two quantities are needed for the calculation of the Kaplan-Meier estimates. Here, for example, we estimate that 79.3% of cases do not experience sex by age 18 years (standard error 6.54%).

Display 10.3 further provides summary information about group sizes and censoring, and estimates of group-wise median latencies together with 95% confidence intervals. The latter estimates (24 years in the cases and 17 years in the controls) take account of the censoring in the data and

```
Survival Analysis for AGESEX   Age of first sex (years)

Factor DIAGNOS = RAN
```

Time	Status	Cumulative Survival	Standard Error	Cumulative Events	Number Remaining
12	Yes	.9750	.0247	1	39
16	Yes			2	38
16	Yes	.9250	.0416	3	37
17	Yes			4	36
17	Yes	.8750	.0523	5	35
17	No			5	34
17	No			5	33
17	No			5	32
18	Yes			6	31
18	Yes			7	30
18	Yes	.7930	.0654	8	29
18	No			8	28
18	No			8	27
18	No			8	26
18	No			8	25
19	Yes			9	24
19	Yes	.7295	.0740	10	23
19	No			10	22
20	Yes			11	21
20	Yes			12	20
20	Yes	.6300	.0833	13	19
20	No			13	18
21	Yes			14	17
21	Yes	.5600	.0875	15	16
21	No			15	15
21	No			15	14
21	No			15	13
22	No			15	12
23	Yes	.5134	.0918	16	11
24	Yes			17	10
24	Yes	.4200	.0959	18	9
24	No			18	8
26	No			18	7
27	No			18	6
28	Yes	.3500	.1024	19	5
28	No			19	4
28	No			19	3
29	No			19	2
30	No			19	1
31	No			19	0

```
Number of Cases:   40        Censored:   21    ( 52.50%)   Events: 19
```

	Survival Time	Standard Error	95% Confidence Interval	
Mean:	24	1	(22,	26)
(Limited to	31)			
Median:	24	2	(20,	28)

Display 10.3   Kaplan-Meier output for survivor function for times to first sex.

differ from naïve group-wise median values of **agesex** which would not acknowledge the censoring (especially in the cases where most censoring occurred). Notably, the two 95% confidence intervals for the median latencies are not overlapping, suggesting group differences in the distributions of age of first sex.

The Kaplan-Meier estimates of the survivor functions of cases and controls become easier to compare if they are both displayed on a single graph. Such a *survival plot* can be requested using the **Options** sub-dialogue box. (Display 10.4, the **Survival table(s)** and the **Mean and median survival** have previously been supplied by default.) The resulting graph is shown in Display 10.5 (after some editing and using the commands **Chart – Reference**

```
Survival Analysis for AGESEX   Age of first sex (years)

Factor DIAGNOS = Control
```

Time	Status	Cumulative Survival	Standard Error	Cumulative Events	Number Remaining
13	Yes	.9773	.0225	1	43
14	Yes	.9545	.0314	2	42
15	Yes			3	41
15	Yes			4	40
15	Yes			5	39
15	Yes	.8636	.0517	6	38
16	Yes			7	37
16	Yes			8	36
16	Yes			9	35
16	Yes			10	34
16	Yes			11	33
16	Yes			12	32
16	Yes			13	31
16	Yes			14	30
16	Yes			15	29
16	Yes			16	28
16	Yes			17	27
16	Yes			18	26
16	Yes			19	25
16	Yes	.5455	.0751	20	24
17	Yes			21	23
17	Yes			22	22
17	Yes			23	21
17	Yes			24	20
17	Yes			25	19
17	Yes			26	18
17	Yes			27	17
17	Yes			28	16
17	Yes	.3409	.0715	29	15
18	Yes			30	14
18	Yes			31	13
18	Yes			32	12
18	Yes			33	11
18	Yes	.2273	.0632	34	10
18	No			34	9
19	Yes			35	8
19	Yes			36	7
19	Yes			37	6
19	Yes			38	5
19	Yes			39	4
19	Yes			40	3
19	Yes	.0505	.0345	41	2
20	Yes			42	1
20	Yes	.0000	.0000	43	0

```
Number of Cases:  44       Censored:   1     ( 2.27%)   Events: 43
```

	Survival Time	Standard Error	95% Confidence Interval	
Mean:	17	0	(16,	17)
Median:	17	0	(16,	18)

```
Survival Analysis for AGESEX   Age of first sex (years)
```

		Total	Number Events	Number Censored	Percent Censored
DIAGNOS	RAN	40	19	21	52.50
DIAGNOS	Control	44	43	1	2.27
Overall		84	62	22	26.19

Display 10.3 (continued)

Line... – Y Scale – Position of Line(s): = 0.5 from the Chart Editor menu to insert a horizontal line at $y = 0.5$).

By definition, both estimated curves start at 1 (100% survival before the first event) and decrease with time. The estimated curves are *step functions* with downward steps occurring at the observed event times. The symbols indicate times when right censoring occurs. The estimated survival curve

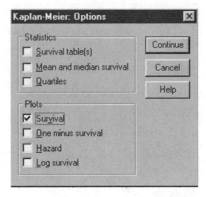

Display 10.4   Requesting a plot of estimated survival curves.

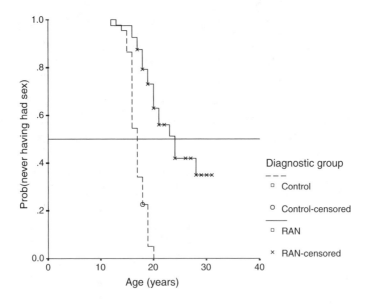

Display 10.5   Estimated survivor functions for times to first sex.

for the control group reaches zero (0% survival) at age 20 years, since by this time all subjects remaining at risk for the event of interest (first sex) had experienced the event. By contrast, the estimated survival curve for the eating disorder group does not reach zero since subjects remain at risk after the largest observed event time (age at first sex 28 years).

The estimated survival curves shown in Display 10.5 demonstrate clearly the older ages at first sex among the cases. At all ages, estimated proportions of survival (i.e., not having had sex) are higher for the cases

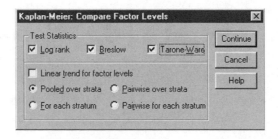

Display 10.6   Requesting tests for comparing survival functions between groups.

than for the controls. The median age in each group is estimated as the time at which the respective survival curve crosses the reference line. Here the controls cross the line later (at 24 years) than the control group (at 17 years).

Although the difference in the survival curves of cases and controls is very clear from Display 10.5, a more formal assessment of this difference is often needed. Survivor functions for different groups of subjects can be compared formally without assuming a particular distribution for the survival times by means of a log-rank test. SPSS supplies the log-rank test and its modified versions, the *generalized Wilcoxon test* (referred to in SPSS as the *Breslow test*, Breslow, 1970) and the *Tarone-Ware test* (Tarone and Ware, 1977). All three tests assess the null hypothesis that the group-wise survival functions do not differ by comparing the observed number of events at each event time with the number expected if the survival experience of the groups were the same. The various versions of the test differ in the differential weights they assign to the event times when constructing the test statistic. The generalized Wilcoxon test uses weights equal to the number at risk and, therefore, puts relatively more weight on differences between the survivor functions at smaller values of time. The log-rank test, which uses weights equal to one at all time points, places more emphasis on larger values of time and the Tarone-Ware test uses a weight function intermediate between those (for more details, see Hosmer and Lemeshow, 1999).

All versions of the log-rank test can be requested via the Compare Factor Levels sub-dialogue box (Display 10.6). Here, for illustrative purposes, we request all three. The resulting output is shown in Display 10.7. The value of the test statistic is largest for the log-rank test, followed by the Tarone-Ware test, and, finally, the generalized Wilcoxon test. Here, this ranking reflects larger differences between the survivor functions at older ages, although the conclusion from each test is the same, namely, that there is very strong evidence for a difference in the survivor functions of the cases and the controls ($X^2(1) = 46.4$, $p < 0.0001$).

```
Survival Analysis for AGESEX    Age of first sex (years)

                            Total      Number       Number      Percent
                                       Events       Censored     Censored

      DIAGNOS    RAN          40         19           21          52.50
      DIAGNOS    Control      44         43            1           2.27

   Overall                    84         62           22          26.19


Test Statistics for Equality of Survival Distributions for DIAGNOS

                     Statistic      df      Significance

   Log Rank           46.37         1          .0000
   Breslow            36.12         1          .0000
   Tarone-Ware        41.16         1          .0000
```

Display 10.7   Log-rank tests output for comparing survival functions for times to first sex between groups.

In the sexual milestones data, only two groups — cases and controls — have to be compared. When a factor of interest has more than two levels, SPSS provides a choice between an overall group test, possibly followed by pairwise group comparisons, and a linear trend test. The default setting in the **Compare Factor Levels** sub-dialogue box provides the overall group test (see Display 10.6). Pairwise group comparisons can be requested by checking **Pairwise over strata**. This provides tests for comparing the survivor functions between any pair of groups. Such comparisons are usually requested post-hoc, i.e., after a significant overall test result has been obtained, and results should be corrected for multiple comparisons. Alternatively, a planned test for trend in survivor functions over the ordered levels of a group factor is available. This test uses a linear polynomial contrast to compare the groups and can be requested by checking **Linear trend for factor levels** and **Pooled over strata** (cf Chapter 5 for more on planned vs. post-hoc comparisons).

Finally, further factors affecting survival but not of direct interest themselves can be adjusted for by carrying out tests for the factor of interest *within* strata defined by various combinations of the levels of the other factors. Such a stratified analysis can be requested by including the stratification variable under the **Strata** list of the **Kaplan-Meier** Dialogue box (see Display 10.2). For any factor level comparison (trend, overall, or all pairwise), the **Compare Factor Levels** sub-dialogue box then provides a choice between carrying out the requested test(s) for each stratum separately (check **For each stratum** or **Pairwise for each stratum** in Display 10.6) or pooling the test(s) over strata (check **Pooled over strata** or **Pairwise over strata**).

10.3.2 WISC Task Completion Times

In the context of the WISC task completion times study, the event of interest is WISC task completion and the survival times are the times taken to complete the task. Children remain at risk until they have completed the task or until the ten minutes allowed has elapsed.

The **Data View** spreadsheet for the WISC task completion times in Table 10.2 contains six variables: a child identifier (**subject**), the time to task completion or time out (**time**), the score on the Embedded Figures Test (**eft**), the age of the child in months (**age**), the experimental group the child was allocated to (**group**), and an event identifier (**status**). The event identifier takes the value "1" when the event of interest, here completion of the WISC task, occurs before the end of the ten-minute test period. If the child ran out of time, its value is "0", indicating **time** is censored. The type of right censoring observed here differs from that in the sexual milestones study in that censoring can only occur at a single time point, namely, the end of the test period (600 sec).

For this study, the main interest is in assessing how the WISC task completion times are affected by the three explanatory variables, **eft**, **age**, and **group**. We will use Cox's regression to investigate this question. As described in Box 10.2, Cox's regression models the hazard function; in this study, this amounts to modeling the rate of immediate task completion at times between 0 and 600 sec.

A Cox regression model can be requested in SPSS using the commands

Analyze – Survival – Cox Regression...

This opens the **Cox Regression** dialogue box shown in Display 10.8. The box is similar to the **Kaplan-Meier** dialogue box (see Display 10.2) and a survival time (**time**) and censoring variable (**status**) need to be declared. In contrast to the latter, the **Cox Regression** dialogue box allows for several explanatory variables to be included under the **Covariates** list. By default, these variables are considered continuous variables. But they can be declared to be factors (and appropriately dummy coded) by using the **Categorical...** button. All our explanatory variables except **group** are continuous. But since this two-level factor has already been coded as a binary dummy variable we were able to include it simply as a covariate. (We also check **CI for exp(B)** and **Display baseline function** in the **Options** sub-dialogue box to generate some additional output.)

(Note the similarity between the Cox and logistic regression dialogue boxes, cf. Display 9.6. In both routines, the **Categorical...** button is used to declare categorical explanatory variables and contrasts of interest and the **>a*b>** button to declare interactions between variables. The Cox regression dialogue also provides a blocking facility and the same options for automatic variable selection procedure as the logistic regression routine. We, therefore, refer the reader to the regression and logistic regression chapters for more details on general model-building strategies, see Chapters 4 and 9.)

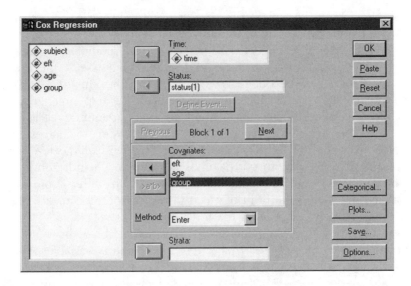

Display 10.8   Declaring a Cox regression model.

The output generated by our Cox regression commands is shown in Display 10.9. The output starts with a "Case Processing Summary" table providing summary information about the censoring in the data set. For example, out of 50 observed times, 26 represented successful task completions. Then, as in logistic regression, SPSS automatically begins by fitting a null model, i.e., a model containing only an intercept parameter. A single table relating to this model, namely the "Omnibus Tests for Model Coefficients" table is provided in the **Output Viewer** under the heading "Block 0: beginning block". The table gives the value of $-2 \times$ Log Likelihood for the null model, which is needed to construct Likelihood ratio (LR) tests for effects of the explanatory variables.

The remainder of the output shown in Display 10.9 is provided in the **Output Viewer** under the heading "Block 1: Method = Enter" and provides details of the requested Cox regression model, here a model containing the covariates **eft, age,** and **group**. The second column of the "Omnibus Tests of Model Coefficients" table (labeled "Overall (score)") provides a score test for simultaneously assessing the effects of the parameters in the model. We find that our three covariates contribute significantly to explaining variability in the WISC task completion hazards (Score test: $X^2(3) = 26.3$, $p < 0.0001$). The remaining columns of the table provide an LR test for comparing the model in the previous output block with this latest one. The change in $-2 \times$ Log Likelihood relative to the null model also indicates a significant improvement in model fit after adding the covariates (LR test: $X^2(3) = 27$, $p < 0.001$). (As in the logistic regression output, an

Case Processing Summary

		N	Percent
Cases available in analysis	Event[a]	26	52.0%
	Censored	24	48.0%
	Total	50	100.0%
Cases dropped	Cases with missing values	0	.0%
	Cases with non-positive time	0	.0%
	Censored cases before the earliest event in astratum	0	.0%
	Total	0	.0%
Total		50	100.0%

a. Dependent Variable: Time to WISC task completion

Omnibus Tests of Model Coefficients

-2 Log Likelihood
187.38 6

Omnibus Tests of Model Coefficients [a,b]

-2 Log Likelihood	Overall (score)			Change From Previous Step			Change From Previous Block		
	Chi-square	df	Sig.	Chi-square	df	Sig.	Chi-square	df	Sig.
160.357	26.304	3	.000	27.029	3	.000	27.029	3	.000

a. Beginning Block Number 0, initial Log Likelihood function: -2 Log likelihood: 187.386
b. Beginning Block Number 1. Method = Enter

Variables in the Equation

	B	SE	Wald	df	Sig.	Exp (B)	95.0% CI for Exp (B)	
							Lower	Upper
EFT	-.115	.025	21.255	1	.000	.892	.849	.936
AGE	-.022	.018	1.450	1	.229	.978	.944	1.014
GROUP	.634	.409	2.403	1	.121	1.885	.846	4.200

Display 10.9   Cox regression output for WISC task completion times.

LR test for the whole block 1 and one for each model building step within the block is given. Since all our three terms were added in a single step within block 1, the two tests are identical.)

The "Variables in the Equation" table provides Wald's tests for assessing the effect of each of the variables in the model after adjusting for the effects of the remaining ones. At the 5% test level, only eft is shown to significantly affect the hazard function after adjusting for the effects of group and age (Wald test: $X^2(1) = 21.3$, $p < 0.001$). The parameter estimates (log-hazard functions) are also given in the column labeled "B", with the column "SE" providing their standard errors. Since effects in terms of log-hazards are hard to interpret, they are generally exponentiated to give hazard ratios (see column headed "Exp(B)"); 95% confidence intervals for the hazard-ratios have been added to the table as requested. The results show that increasing the EFT score by one point decreases the hazard function by 10.8% (95% CI from 6.4 to 15.1%). Or, if it is thought more appropriate to express the effect of EFT as that of a 10 point increase since the scores in our sample varied between 15 and 63, then the hazard

Survival Table

Time	Baseline Cum Hazard	At mean of covariates		
		Survival	SE	Cum Hazard
121.750	6.980	.992	.008	.008
147.430	14.612	.982	.013	.018
214.880	23.977	.971	.018	.029
223.780	35.700	.958	.023	.043
258.700	48.148	.943	.028	.058
272.820	61.256	.928	.032	.074
290.080	75.123	.913	.036	.091
339.610	90.381	.896	.040	.109
359.790	106.629	.879	.044	.129
371.950	123.416	.861	.048	.150
385.310	140.897	.843	.051	.171
389.140	159.922	.824	.055	.194
389.330	180.864	.803	.058	.219
396.130	202.352	.783	.061	.245
403.610	224.116	.762	.064	.272
429.580	246.349	.742	.066	.298
461.610	269.192	.722	.069	.326
462.140	294.233	.700	.071	.356
472.210	321.370	.677	.073	.389
473.150	349.949	.654	.075	.424
490.610	381.003	.630	.076	.462
530.180	413.692	.606	.078	.501
545.720	447.572	.581	.079	.542
559.250	483.626	.557	.080	.586
563.550	524.743	.530	.080	.636
589.250	570.167	.501	.081	.691

Covariate Means

	Mean
EFT	41.841
AGE	100.865
GROUP	.500

Display 10.9 (continued)

of task completion decreases by 68.3% (95% CI from 48.3 to 80.5%). Clearly, children with a greater field-dependent cognitive style take longer to complete the WISC task.

The "Survival Table" provides the estimated baseline cumulative hazard function for WISC task completion. Also given are the resulting predicted survivor function and predicted cumulative hazard function at the mean values of the covariates. The fitted baseline hazard of the Cox model is not usually of interest. However, the gradient of the cumulative hazard curve informs us about the hazard at a given time point. Here we notice that the hazard of immediate task completion increases somewhat in the second half of the test period.

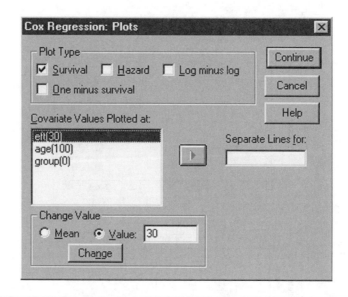

Display 10.10 **Plotting the predicted survival curve for a set of covariate values.**

It is often helpful to graph the fitted Cox model in some way. Here, we will try to illustrate the effect of EFT on the predicted survivor function. To do this, we display the predicted survival curves for fixed values of the covariates **group** (value 0 = row group) and **age** (value 100 months) but varying values of the covariate **eft** (EFT scores of 30, 40, and 50). The predicted survival curve for a set of covariate values can be plotted in SPSS by using the **Plots** sub-dialogue box (Display 10.10). By default, the predicted survival curve is plotted at the mean values of the covariates, but the values can be changed as indicated in Display 10.10. (SPSS can display survivor curves for each level of a factor in the same plot using the **Separate Lines for** facility. However, this facility is limited to categorical explanatory variables, so here we have to repeat the procedure three times, each time generating a survival curve at a single value of EFT.)

The resulting plots are shown in Display 10.11. We have enhanced the plots by including grid lines for easier comparison across plots (by double-clicking on the axes in the **Chart Editor** and requesting **Grid for Minor Divisions**). Comparing curves a) to c), the marked increase in survival probabilities (task noncompletion probabilities) at all times for an extra 10 points EFT is apparent.

Cox's regression is a semi-parametric procedure and as such does not assume that the functional form of the survival time distribution is known. However, the procedure relies on the proportional hazards assumption and an assumption of linearity (see Box 10.2). The proportional hazards

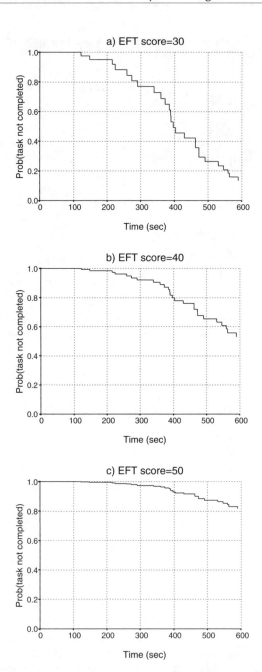

Display 10.11   Predicted survival curves for WISC task completion times at different values of the EFT covariate. The covariates group and age have been held constant at values 0 (row group) and 100 (months age), respectively.

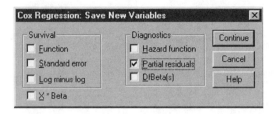

Display 10.12   Saving results from a Cox regression.

assumption is expected to hold for each of the covariates included in the model. For example, for the binary covariate **group**, the assumption states that the hazard ratio of completing the task between a child in the row group and one in the corner group is constant over time. For continuous covariates such as EFT, the assumption states that proportionality of hazard functions holds for any two levels of the covariate. In addition, accommodating continuous covariates requires the linearity assumption. This states that their effect is linear on the log-hazard scale.

Schoenfeld residuals (Schoenfeld, 1982) can be employed to check for proportional hazards. The residuals can be constructed for each covariate and contain values for subjects with uncensored survival times. Under the proportional hazard assumption for the respective covariate, a scatterplot of Schoenfeld residuals against event times (or log-times) is expected to scatter in a nonsystematic way about the zero line, and the polygon connecting the values of the smoothed residuals should have approximately a zero slope and cross the zero line several times (for more on residual diagnostics, see Hosmer and Lemeshow, 1999).

SPSS refers to the Schoenfeld residuals as "partial residuals" and supplies them via the **Save...** button on the **Cox Regression** dialogue box. The resulting sub-dialogue box is shown in Display 10.12. It allows saving information regarding the predicted survival probability at the covariate values of each case and three types of residual diagnostics. We opt for the Schoenfeld residuals. This results in the inclusion of three new variables, labeled **pr1_1** ("partial" residual for **eft**), **pr2_1** (residual for **age**), and **pr3_1** (residual for **group**) in our **Data View** spreadsheet. We then proceed to plot these residuals against task completion times (**time**). The resulting scatterplots become more useful if we include a Lowess curve of the residuals and a horizontal reference line that crosses the *y*-axis at its origin.

The resulting residual plots are shown in Display 10.13. The residual plot for EFT indicates no evidence of a departure from the proportional hazards assumption for this covariate. As expected, the Lowess curve varies around the zero reference line in a nonsystematic way (Display 10.13a). For covariates **age** and **group**, the situation is less clear.

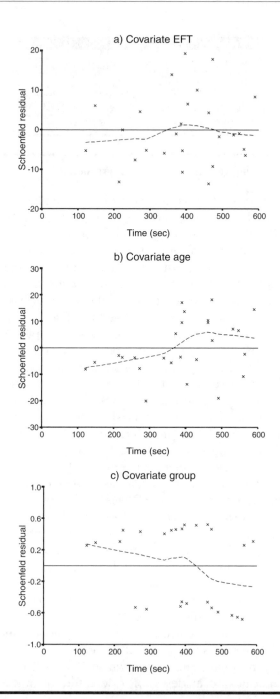

Display 10.13 Schoenfeld residual plots for the covariates in the Cox regression model for WISC task completion times.

The smooth lines in the respective residual plots appear to demonstrate a systematic trend over time (Displays 10.13b and c). For example, for **group** the polygon appears to have a negative slope, suggesting that the hazard ratio of the covariate decreases over time and thus that it has a nonproportional hazard. However, the plots are only based on 26 events and, at least for age, the trend is largely influenced by the residuals at the first two time points. (Note also the two bands of residuals for the binary covariate. The upper band corresponds to children with **group** = 1, i.e., the corner group, and the lower band to those with **group** = 0.)

Another type of residual used in Cox models is the *Martingale residual* (Therneau et al., 1990). These residuals cannot be saved directly by the **Save New variables** sub-dialogue box, but they can easily be generated from *Cox-Snell residuals* (Cox and Snell, 1968), which are just the estimated cumulative hazards, and then saved. For a censored case, the Martingale residual is the negative of the Cox-Snell residual. For an uncensored case, it is one minus the Cox-Snell residual.

Covariate-specific Martingale residuals can be used to assess the linearity assumption. They are constructed by generating residuals from Cox models that set the effect of the covariate in question to zero. These Martingale residuals can then be plotted against the respective covariate to indicate the functional form of the relationship between the log-hazard function and the covariate.

We can generate Martingale residuals for the continuous covariates EFT and age using the following series of steps:

- Exclude the covariate in question from the **Covariates** list on the **Cox Regression** dialogue box (see Display 10.8).
- Check **Hazard function** on the **Save New Variables** sub-dialogue box (see Display 10.12).
- Run the Cox regressions to add Cox-Snell residuals to the **Data View** spreadsheet (labeled **haz_1** and **haz_2**).
- Use the **Compute** command to calculate Martingale residuals, **mar_1** and **mar_2** from the formulae **mar_1 = status-haz_1** and **mar_2 = status-haz_2**.

We can now plot the Martingale residuals against the respective covariates and enhance the plots by including Lowess curves. The resulting plots are shown in Display 10.14. For age, assuming a linear relationship seems reasonable. In fact, the line is almost constant over the levels of the covariate, indicating that the hazard of task completion is little affected by age. A linear function also seems a reasonable approximation of the relationships between the log-hazards and EFT, although, perhaps, the chance of task completion reaches a minimum value at EFT scores around 50.

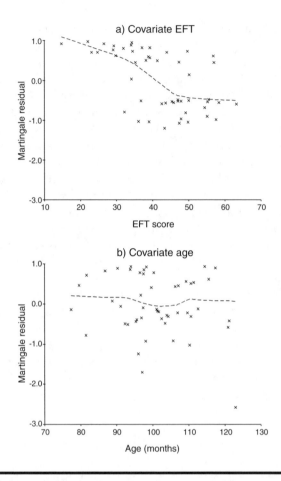

Display 10.14 Martingale residual plots for the continuous covariates in the Cox regression model for WISC task completion times.

Since some doubt has been shed on the proportional hazards assumption for the covariate **group** and also because our primary aim was to predict WISC task completion times rather than formally compare groups, we now show how to fit a Cox's regression *stratified* by our experimental groups. This approach allows for the possibility of there being different baseline hazards in the two experimental groups and so avoids the proportional hazards assumption.

This stratified Cox model can be declared in SPSS by removing the stratification variable, **group**, from the **Covariates** list and instead including it under the **Strata** list of the **Cox Regression** dialogue box (see Display 10.8).

The SPSS output from a stratified Cox regression looks similar to that of an unstratified analysis, except that separate baseline hazard functions are estimated for each level of the stratifying variable. Comparison of the

Stratum Status[a]

Stratum	Strata label	Event	Censored	Censored Percent
0	Row group	11	14	56.0%
1	Corner group	15	10	40.0%
Total		26	24	48.0%

[a.] The strata variable is: GROUP

Variables in the Equation

	B	SE	Wald	df	Sig.	Exp(B)	95.0% CI for Exp(B) Lower	95.0% CI for Exp(B) Upper
EFT	-.115	.025	20.932	1	.000	.892	.849	.937
AGE	-.022	.019	1.386	1	.239	.978	.943	1.015

Display 10.15 Selected output from a stratified Cox regression of the WISC task completion times.

regression parameter output from the stratified model (Display 10.15) with that from the unstratified model (see Display 10.9) shows that the effects of EFT and age are almost unchanged. We can, therefore, be relatively confident that the immediate WISC task completion rates are predicted by the EFT variable with the rate of completion at any time point decreasing by 10.8% for every additional point on the EFT (95% CI from 6.3 to 15.1%).

10.4 Exercises

10.4.1 Gastric Cancer

The data below show the survival times in days of two groups of 45 patients suffering from gastric cancer (data given in Gamerman, 1991). Group 1 received chemotherapy and radiation; group 2 received only chemotherapy. An asterisk indicates censoring. Plot the Kaplan-Meier survival curves of each group and test for a difference using the log-rank test.

Group 1

17, 42, 44, 48, 60, 72, 74, 95, 103, 108, 122, 144, 167, 170, 183, 185, 193, 195, 197, 208, 234, 235, 254, 307, 315, 401, 445, 464, 484, 528, 542, 567, 577, 580, 795, 855, 1174*, 1214, 1232*, 1366, 1455*, 1585*, 1622*, 1626*, 1936*

Group2

1, 63, 105, 125, 182, 216, 250, 262, 301, 342, 354, 356, 358, 380, 383, 383, 388, 394, 408, 460, 489, 523, 524, 535, 562, 569, 675, 676, 748, 778, 786, 797, 955, 968, 977, 1245, 1271, 1420, 1460*, 1516*, 1551, 1690*, 1694

10.4.2 Heroin Addicts

The data shown in Table 10.3 give the times that heroin addicts remained in a clinic for methadone maintenance treatment. Here the endpoint of interest is not death, but termination of treatment. Some subjects were still in the clinic at the time these data were recorded or failed to complete their treatment program (status variable: "1" for subjects who had departed the clinic on completion of treatment and "0" otherwise). Possible explanatory variables for predicting time to complete treatment are maximum methadone dose, whether or not the addict had a criminal record, and the clinic in which the addict was treated. Fit a Cox's regression model to these data and consider the possibility of stratifying on clinic.

10.4.3 More on Sexual Milestones of Females

The times to first sexual intercourse shown in Table 10.1 were analyzed using a nonparametric survival analysis in the main body of the text. If appropriate, reanalyze this data set using a semi-parametric survival analysis. Specifically:

- Assess the proportional hazards assumption. (Hint: Under the proportional hazards assumption, Kaplan-Meier estimates of the cumulative hazard function in the two groups are expected to be proportional.)
- Fit a Cox regression and interpret your findings.
- Plot the survivor functions and cumulative hazard functions predicted by the fitted Cox model and compare these plots with Kaplan-Meier estimates of the curves.

Table 10.3   Heroin Addicts Data

Subject	Clinic	Status	Time	Prison?	Dose	Subject	Clinic	Status	Time	Prison?	Dose
1	1	1	428	0	50	132	2	0	633	0	70
2	1	1	275	1	55	133	2	1	661	0	40
3	1	1	262	0	55	134	2	1	232	1	70
4	1	1	183	0	30	135	2	1	13	1	60
5	1	1	259	1	65	137	2	0	563	0	70
6	1	1	714	0	55	138	2	0	969	0	80
7	1	1	438	1	65	143	2	0	1052	0	80
8	1	0	796	1	60	144	2	0	944	1	80
9	1	1	892	0	50	145	2	0	881	0	80
10	1	1	393	1	65	146	2	1	190	1	50
11	1	0	161	1	80	148	2	1	79	0	40
12	1	1	836	1	60	149	2	0	884	1	50
13	1	1	523	0	55	150	2	1	170	0	40
14	1	1	612	0	70	153	2	1	286	0	45
15	1	1	212	1	60	156	2	0	358	0	60
16	1	1	399	1	60	158	2	0	326	1	60
17	1	1	771	1	75	159	2	0	769	1	40
18	1	1	514	1	80	160	2	1	161	0	40
19	1	1	512	0	80	161	2	0	564	1	80
21	1	1	624	1	80	162	2	1	268	1	70
22	1	1	209	1	60	163	2	0	611	1	40
23	1	1	341	1	60	164	2	1	322	0	55
24	1	1	299	0	55	165	2	0	1076	1	80
25	1	0	826	0	80	166	2	0	2	1	40
26	1	1	262	1	65	168	2	0	788	0	70
27	1	0	566	1	45	169	2	0	575	0	80
28	1	1	368	1	55	170	2	1	109	1	70
30	1	1	302	1	50	171	2	0	730	1	80
31	1	0	602	0	60	172	2	0	790	0	90
32	1	1	652	0	80	173	2	0	456	1	70
33	1	1	293	0	65	175	2	1	231	1	60
34	1	0	564	0	60	176	2	1	143	1	70
36	1	1	394	1	55	177	2	0	86	1	40
37	1	1	755	1	65	178	2	0	1021	0	80
38	1	1	591	0	55	179	2	0	684	1	80
39	1	0	787	0	80	180	2	1	878	1	60
40	1	1	739	0	60	181	2	1	216	0	100
41	1	1	550	1	60	182	2	0	808	0	60
42	1	1	837	0	60	183	2	1	268	1	40
43	1	1	612	0	65	184	2	0	222	0	40
44	1	0	581	0	70	186	2	0	683	0	100

Table 10.3 (continued)   Heroin Addicts Data

Subject	Clinic	Status	Time	Prison?	Dose	Subject	Clinic	Status	Time	Prison?	Dose
45	1	1	523	0	60	187	2	0	496	0	40
46	1	1	504	1	60	188	2	1	389	0	55
48	1	1	785	1	80	189	1	1	126	1	75
49	1	1	774	1	65	190	1	1	17	1	40
50	1	1	560	0	65	192	1	1	350	0	60
51	1	1	160	0	35	193	2	0	531	1	65
52	1	1	482	0	30	194	1	0	317	1	50
53	1	1	518	0	65	195	1	0	461	1	75
54	1	1	683	0	50	196	1	1	37	0	60
55	1	1	147	0	65	197	1	1	167	1	55
57	1	1	563	1	70	198	1	1	358	0	45
58	1	1	646	1	60	199	1	1	49	0	60
59	1	1	899	0	60	200	1	1	457	1	40
60	1	1	857	0	60	201	1	1	127	0	20
61	1	1	180	1	70	202	1	1	7	1	40
62	1	1	452	0	60	203	1	1	29	1	60
63	1	1	760	0	60	204	1	1	62	0	40
64	1	1	496	0	65	205	1	0	150	1	60
65	1	1	258	1	40	206	1	1	223	1	40
66	1	1	181	1	60	207	1	0	129	1	40
67	1	1	386	0	60	208	1	0	204	1	65
68	1	0	439	0	80	209	1	1	129	1	50
69	1	0	563	0	75	210	1	1	581	0	65
70	1	1	337	0	65	211	1	1	176	0	55
71	1	0	613	1	60	212	1	1	30	0	60
72	1	1	192	1	80	213	1	1	41	0	60
73	1	0	405	0	80	214	1	0	543	0	40
74	1	1	667	0	50	215	1	0	210	1	50
75	1	0	905	0	80	216	1	1	193	1	70
76	1	1	247	0	70	217	1	1	434	0	55
77	1	1	821	0	80	218	1	1	367	0	45
78	1	1	821	1	75	219	1	1	348	1	60
79	1	0	517	0	45	220	1	0	28	0	50
80	1	0	346	1	60	221	1	0	337	0	40
81	1	1	294	0	65	222	1	0	175	1	60
82	1	1	244	1	60	223	2	1	149	1	80
83	1	1	95	1	60	224	1	1	546	1	50
84	1	1	376	1	55	225	1	1	84	0	45
85	1	1	212	0	40	226	1	0	283	1	80
86	1	1	96	0	70	227	1	1	533	0	55

Table 10.3 (continued)   Heroin Addicts Data

Subject	Clinic	Status	Time	Prison?	Dose	Subject	Clinic	Status	Time	Prison?	Dose
87	1	1	532	0	80	228	1	1	207	1	50
88	1	1	522	1	70	229	1	1	216	0	50
89	1	1	679	0	35	230	1	0	28	0	50
90	1	0	408	0	50	231	1	1	67	1	50
91	1	0	840	0	80	232	1	0	62	1	60
92	1	0	148	1	65	233	1	0	111	0	55
93	1	1	168	0	65	234	1	1	257	1	60
94	1	1	489	0	80	235	1	1	136	1	55
95	1	0	541	0	80	236	1	0	342	0	60
96	1	1	205	0	50	237	2	1	41	0	40
97	1	0	475	1	75	238	2	0	531	1	45
98	1	1	237	0	45	239	1	0	98	0	40
99	1	1	517	0	70	240	1	1	145	1	55
100	1	1	749	0	70	241	1	1	50	0	50
101	1	1	150	1	80	242	1	0	53	0	50
102	1	1	465	0	65	243	1	0	103	1	50
103	2	1	708	1	60	244	1	0	2	1	60
104	2	0	713	0	50	245	1	1	157	1	60
105	2	0	146	0	50	246	1	1	75	1	55
106	2	1	450	0	55	247	1	1	19	1	40
109	2	0	555	0	80	248	1	1	35	0	60
110	2	1	460	0	50	249	2	0	394	1	80
111	2	0	53	1	60	250	1	1	117	0	40
113	2	1	122	1	60	251	1	1	175	1	60
114	2	1	35	1	40	252	1	1	180	1	60
118	2	0	532	0	70	253	1	1	314	0	70
119	2	0	684	0	65	254	1	0	480	0	50
120	2	0	769	1	70	255	1	0	325	1	60
121	2	0	591	0	70	256	2	1	280	0	90
122	2	0	769	1	40	257	1	1	204	0	50
123	2	0	609	1	100	258	2	1	366	0	55
124	2	0	932	1	80	259	2	0	531	1	50
125	2	0	932	1	80	260	1	1	59	1	45
126	2	0	587	0	110	261	1	1	33	1	60
127	2	1	26	0	40	262	2	1	540	0	80
128	2	0	72	1	40	263	2	0	551	0	65
129	2	0	641	0	70	264	1	1	90	0	40
131	2	0	367	0	70	266	1	1	47	0	45

Source: Hand et al., 1994.

Chapter 11

Principal Component Analysis and Factor Analysis: Crime in the U.S. and AIDS Patients' Evaluations of Their Clinicians

11.1 Description of Data

In this chapter, we are concerned with two data sets; the first shown in Table 11.1 gives rates of various types of crime in the states of the U.S. The second data set, shown in Table 11.2, arises from a survey of AIDS patients' evaluations of their physicians. The 14 items in the survey questionnaire measure patients' attitudes toward a clinician's personality, demeanour, competence, and prescribed treatment; each item is measured using a Likert-type scale ranging from one to five. Since seven of the items are stated negatively, they have been recoded (reflected) so that one always represents the most positive attitude and five the least positive. The 14 items are described in Table 11.2.

Table 11.1   Crime Rates in the U.S.[a]

State	Murder	Rape	Robbery	Aggravated Assault	Burglary	Larceny/Theft	Motor Vehicle Theft
ME	2	14.8	28	102	803	2347	164
NH	2.2	21.5	24	92	755	2208	228
VT	2	21.8	22	103	949	2697	181
MA	3.6	29.7	193	331	1071	2189	906
RI	3.5	21.4	119	192	1294	2568	705
CT	4.6	23.8	192	205	1198	2758	447
NY	10.7	30.5	514	431	1221	2924	637
NJ	5.2	33.2	269	265	1071	2822	776
PA	5.5	25.1	152	176	735	1654	354
OH	5.5	38.6	142	235	988	2574	376
IN	6	25.9	90	186	887	2333	328
IL	8.9	32.4	325	434	1180	2938	628
MI	11.3	67.4	301	424	1509	3378	800
WI	3.1	20.1	73	162	783	2802	254
MN	2.5	31.8	102	148	1004	2785	288
IA	1.8	12.5	42	179	956	2801	158
MO	9.2	29.2	170	370	1136	2500	439
ND	1	11.6	7	32	385	2049	120
SD	4	17.7	16	87	554	1939	99
NE	3.1	24.6	51	184	784	2677	168
KS	4.4	32.9	80	252	1188	3008	258
DE	4.9	56.9	124	241	1042	3090	272
MD	9	43.6	304	476	1296	2978	545
DC	31	52.4	754	668	1728	4131	975
VA	7.1	26.5	106	167	813	2522	219
WV	5.9	18.9	41	99	625	1358	169
NC	8.1	26.4	88	354	1225	2423	208
SC	8.6	41.3	99	525	1340	2846	277
GA	11.2	43.9	214	319	1453	2984	430
FL	11.7	52.7	367	605	2221	4372	598
KY	6.7	23.1	83	222	824	1740	193
TN	10.4	47.	208	274	1325	2126	544
AL	10.1	28.4	112	408	1159	2304	267
MS	11.2	25.8	65	172	1076	1845	150
AR	8.1	28.9	80	278	1030	2305	195
LA	12.8	40.1	224	482	1461	3417	442
OK	8.1	36.4	107	285	1787	3142	649
TX	13.5	51.6	240	354	2049	3987	714
MT	2.9	17.3	20	118	783	3314	215
ID	3.2	20	21	178	1003	2800	181

Table 11.1 (continued)   Crime Rates in the U.S.ᵃ

State	Murder	Rape	Robbery	Aggravated Assault	Burglary	Larceny/Theft	Motor Vehicle Theft
WY	5.3	21.9	22	243	817	3078	169
CO	7	42.3	145	329	1792	4231	486
NM	11.5	46.9	130	538	1846	3712	343
AZ	9.3	43	169	437	1908	4337	419
UT	3.2	25.3	59	180	915	4074	223
NV	12.6	64.9	287	354	1604	3489	478
WA	5	53.4	135	244	1861	4267	315
OR	6.6	51.1	206	286	1967	4163	402
CA	11.3	44.9	343	521	1696	3384	762
AK	8.6	72.7	88	401	1162	3910	604
HI	4.8	31	106	103	1339	3759	328

ᵃ Data are the number of offenses known to police per 100,000 residents of 50 states plus the District of Columbia for the year 1986.

Source: Statistical Abstract of the USA, 1988, Table 265.

For the crime data, the main aim of our analysis is to try to uncover any interesting patterns of crime among the different states. For the AIDS data, we explore whether there is a relatively simple underlying structure that produces the observed associations among the 14 questionnaire items.

11.2 Principal Component and Factor Analysis

Two methods of analysis are the subject of this chapter, *principal component analysis* and *factor analysis*. In very general terms, both can be seen as approaches to summarizing and uncovering any patterns in a set of multivariate data, essentially by reducing the complexity of the data. The details behind each method are quite different.

11.2.1 Principal Component Analysis

Principal component analysis is a multivariate technique for transforming a set of related (correlated) variables into a set of unrelated (uncorrelated) variables that account for decreasing proportions of the variation of the original observations. The rationale behind the method is an attempt to reduce the complexity of the data by decreasing the number of variables

Table 11.2 AIDS Patient's Evaluation of Clinician

Q1   My doctor treats me in a friendly manner.
Q2   I have some doubts about the ability of my doctor.
Q3   My doctor seems cold and impersonal.
Q4   My doctor does his/her best to keep me from worrying.
Q5   My doctor examines me as carefully as necessary.
Q6   My doctor should treat me with more respect.
Q7   I have some doubts about the treatment suggested by my doctor.
Q8   My doctor seems very competent and well trained.
Q9   My doctor seems to have a genuine interest in me as a person.
Q10 My doctor leaves me with many unanswered questions about my condition and its treatment.
Q11 My doctor uses words that I do not understand.
Q12 I have a great deal of confidence in my doctor.
Q13 I feel I can tell my doctor about very personal problems.

Q1	Q2	Q3	Q4	Q5	Q6	Q7	Q8	Q9	Q10	Q11	Q12	Q13	Q14
1	2	2	3	2	2	4	1	1	2	2	1	2	2
2	3	2	3	2	2	3	2	2	3	3	2	1	2
2	2	3	3	4	4	2	2	4	2	2	4	4	2
2	4	2	2	2	4	4	4	2	4	4	4	4	2
2	2	2	2	2	2	2	2	2	2	2	2	2	2
2	4	2	2	2	2	2	1	2	2	2	2	2	1
2	2	1	5	2	1	1	1	1	2	4	2	2	2
2	2	2	2	2	2	2	2	2	2	2	2	2	5
1	1	2	2	2	1	2	1	1	2	2	2	2	1
2	3	2	3	2	2	3	3	3	4	4	3	2	2
2	2	2	2	2	2	4	2	2	2	2	2	2	2
2	1	4	1	2	1	1	1	3	1	1	1	2	1
2	2	2	2	2	2	2	2	2	2	2	2	2	2
5	2	2	5	4	5	4	3	5	5	2	3	4	4
2	2	2	4	2	3	4	2	2	4	2	2	2	2
1	1	1	1	1	1	2	2	1	1	1	1	1	2
1	1	1	1	1	1	1	1	1	2	3	1	1	1
1	1	2	1	1	2	1	1	1	1	4	1	2	1
1	1	1	1	1	1	3	1	1	1	2	1	1	1
1	1	1	1	1	2	3	1	1	1	1	1	1	1
2	3	1	3	1	2	3	2	3	4	2	2	2	3
1	2	1	1	3	1	3	1	1	3	1	1	1	1
2	2	2	2	2	2	2	2	2	2	2	2	2	2
2	3	3	4	5	3	4	3	4	5	4	4	2	2
1	1	2	1	1	1	1	1	2	1	5	1	1	1

Table 11.2 (continued)   AIDS Patient's Evaluation of Clinician

Q1	Q2	Q3	Q4	Q5	Q6	Q7	Q8	Q9	Q10	Q11	Q12	Q13	Q14
2	3	2	2	2	2	4	2	2	4	4	3	2	3
1	2	1	1	1	2	3	1	1	2	2	2	2	2
1	1	1	1	1	2	2	1	1	2	2	2	1	1
1	1	2	2	1	2	1	1	2	4	1	1	2	1
2	1	1	2	2	1	2	1	2	2	4	2	2	2
1	1	1	2	2	2	2	2	2	2	2	2	2	2
1	1	1	3	3	1	1	1	1	2	1	1	1	1
1	1	1	1	1	1	1	1	1	1	3	1	1	1
2	1	2	2	2	2	2	1	3	2	2	2	3	2
1	1	1	2	2	2	3	2	1	1	4	2	2	1
2	2	2	2	2	2	2	2	2	2	2	3	2	2
4	4	4	4	2	4	2	3	4	2	4	2	3	2
1	1	1	2	2	2	1	1	1	1	4	1	1	1
2	1	1	2	1	3	4	1	1	3	4	2	2	2
2	3	4	2	2	2	2	1	2	2	5	3	2	2
1	1	1	1	1	1	2	1	2	2	2	1	2	2
2	4	3	2	2	2	4	2	2	4	3	2	2	1
1	2	2	1	3	2	2	2	2	2	1	2	2	1
1	3	2	2	2	2	2	2	2	2	2	2	1	2
1	1	1	4	1	1	1	1	1	1	1	1	1	1
1	1	1	1	1	1	1	1	1	1	1	1	1	1
1	1	1	1	1	1	1	1	1	1	2	1	1	1
2	2	2	2	3	4	4	2	2	4	4	2	2	5
1	2	1	2	3	2	2	2	2	2	2	2	2	1
4	4	5	4	4	4	4	3	4	5	4	4	5	4
2	1	1	2	2	2	2	2	2	1	2	1	1	2
2	2	2	2	2	2	2	2	2	2	4	2	2	2
2	2	2	2	2	4	4	2	2	2	2	2	2	2
2	2	2	2	2	2	2	2	2	2	4	2	2	2
2	2	2	2	2	2	2	2	2	2	3	4	2	2
2	2	2	2	2	2	2	2	2	2	2	2	2	2
1	1	1	1	1	2	2	1	1	1	2	1	1	1
2	4	2	2	3	2	2	4	4	4	2	5	1	2
2	3	2	2	2	2	2	2	2	2	2	2	2	2
2	4	2	4	2	3	3	2	3	4	2	4	4	4
1	1	1	1	1	1	1	1	1	1	1	1	1	1
1	1	1	2	1	1	1	1	1	2	2	1	1	1
1	2	1	2	2	2	1	1	1	1	1	2	2	1
1	1	1	1	2	1	1	1	1	2	2	1	1	1

that need to be considered. If the first few of the derived variables (the *principal components*) among them account for a large proportion of the total variance of the observed variables, they can be used both to provide a convenient summary of the data and to simplify subsequent analyses.

The coefficients defining the principal components are found by solving a series of equations involving the elements of the observed covariance matrix, although when the original variables are on very different scales, it is wiser to extract them from the observed correlation matrix instead. (An outline of principal component analysis is given in Box 11.1.)

Choosing the number of components to adequately summarize a set of multivariate data is generally based on one of a number of relative ad hoc procedures:

■ Retain just enough components to explain some specified large percentages of the total variation of the original variables. Values between 70 and 90% are usually suggested, although smaller values might be appropriate as the number of variables, q, or number of subjects, n, increases.

■ Exclude those principal components with variances less than the average. When the components are extracted from the observed correlation matrix, this implies excluding components with variances less than one. (This is a very popular approach but has its critics; see, for example, Preacher and MacCallam, 2003.)

■ Plot the component variances as a *scree diagram* and look for a clear "elbow" in the curve (an example will be given later in the chapter).

Often, one of the most useful features of the results of a principal components analysis is that it can be used to construct an informative graphical display of multivariate data that may assist in understanding the structure of the data. For example, the first two principal component scores for each sample member can be plotted to produce a scatterplot of the data. If more than two components are thought necessary to adequately represent the data, other component scores can be used in three-dimensional plots or in scatterplot matrices.

Box 11.1 *Principal Component Analysis*

■ Principal components is essentially a method of data reduction that aims to produce a small number of derived variables that can be used in place of the larger number of original variables to simplify subsequent analysis of the data.

- The principal component variables y_1, y_2, ..., y_q are defined to be linear combinations of the original variables x_1, x_2, ..., x_q that are uncorrelated and account for maximal proportions of the variation in the original data, i.e., y_1 accounts for the maximum amount of the variance among all possible linear combinations of x_1, ..., x_q, y_2 accounts for the maximum variance subject to being uncorrelated with y_1 and so on.
- Explicitly, the principal component variables are obtained from x_1, ..., x_q as follows:

$$y_1 = a_{11}x_1 + a_{12}x_2 + \cdots + a_{1q}x_q$$

$$y_2 = a_{21}x_1 + a_{22}x_2 + \cdots + a_{2q}x_q$$

$$\vdots$$

$$y_q = a_{q1}x_1 + a_{q2}x_2 + \cdots + a_{qq}x_q$$

where the coefficients a_{ij} ($i = 1, ..., q, j = 1, ..., q$) are chosen so that the required maximal variance and uncorrelated conditions hold.
- Since the variances of the principal components variables could be increased without limit, simply by increasing the coefficients that define them, a restriction must be placed on these coefficients.
- The constraint usually applied is that the sum of squares of the coefficients is one so that the total variance of all the components is equal to the total variance of all the observed variables.
- It is often convenient to rescale the coefficients so that their sum of squares are equal to the variance of that component they define. In the case of components derived from the correlation matrix of the data, these rescaled coefficients give the correlations between the components and the original variables. It is these values that are often presented as the result of a principal components analysis.
- The coefficients defining the principal components are given by what are known as the *eigenvectors* of the sample covariance matrix, **S**, or the correlation matrix, **R**. Components derived from **S** may differ considerably from those derived from **R**, and there is not necessarily any simple relationship between them. In most practical applications of principal components,

the analysis is based on the correlation matrix, i.e., on the standardized variables, since the original variables are likely to be on very different scales so that linear combinations of them will make little sense.

■ Principal component scores for an individual i with vector of variable values $\mathbf{x}_i^T$ can be obtained by simply applying the derived coefficients to the observed variables, generally after subtracting the mean of the variable, i.e., from the equations

$$y_{i1} = \mathbf{a}_1^T\left(\mathbf{x}_i - \overline{\mathbf{x}}\right)$$

$$\vdots$$

$$y_{iq} = \mathbf{a}_q^T\left(\mathbf{x}_i - \overline{\mathbf{x}}\right)$$

where $\mathbf{a}_i^T = [a_{i1}, a_{i2}, \cdots, a_{iq}]$, and $\overline{\mathbf{x}}$ is the mean vector of the observations.

(Full details of principal components analysis are given in Everitt and Dunn, 2001, and Jolliffe, 2002.)

11.2.2 Factor Analysis

Factor analysis (more properly *exploratory factor analysis*) is concerned with whether the covariances or correlations between a set of observed variables can be explained in terms of a smaller number of unobservable constructs known either as *latent variables* or *common factors*. Explanation here means that the correlation between each pair of measured (*manifest*) variables arises because of their mutual association with the common factors. Consequently, the partial correlations between any pair of observed variables, given the values of the common factors, should be approximately zero.

Application of factor analysis involves the following two stages:

■ Determining the number of common factors needed to adequately describe the correlations between the observed variables, and estimating how each factor is related to each observed variable (i.e., estimating the *factor loadings*);
■ Trying to simplify the initial solution by the process known as *factor rotation*.

(Exploratory factor analysis is outlined in Box 11.2.)

Box 11.2 Exploratory Factor Analysis

- In general terms, exploratory factor analysis is concerned with whether the covariances or correlations between a set of observed variables $x_1, x_2, \ldots, x_q$ can be 'explained' in terms of a smaller number of unobservable latent variables or common factors, $f_1, f_2, \ldots, f_k$, where $k < q$ and hopefully much less.
- The formal model linking manifest and latent variables is simply that of multiple regression (see Chapter 4), with each observed variable being regressed on the common factors. The regression coefficients in the model are known in this context as the *factor loadings* and the random error terms as *specific variates* since they now represent that part of an observed variable not accounted for by the common factors.
- In mathematical terms, the factor analysis model can be written as follows:

$$x_1 = \lambda_{11} f_1 + \lambda_{12} f_2 \cdots + \lambda_{1k} f_k + u_1$$

$$x_2 = \lambda_{21} f_1 + \lambda_{22} f_2 \cdots + \lambda_{2k} f_k + u_2$$

$$\vdots$$

$$x_q = \lambda_{q1} f_1 + \lambda_{q1} f_2 \cdots + \lambda_{qk} f_k + u_q$$

- The equations above can be written more concisely as

$$\mathbf{x} = \Lambda \mathbf{f} + \mathbf{u}$$

where

$$\mathbf{x} = \begin{bmatrix} x_1 \\ \vdots \\ x_q \end{bmatrix}, \quad \Lambda = \begin{bmatrix} \lambda_{11} & \cdots & \lambda_{1k} \\ \vdots & & \vdots \\ \lambda_{q1} & \cdots & \lambda_{qk} \end{bmatrix}, \quad \mathbf{f} = \begin{bmatrix} f_1 \\ \vdots \\ f_k \end{bmatrix}, \quad \mathbf{u} = \begin{bmatrix} u_1 \\ \vdots \\ u_q \end{bmatrix}$$

- Since the factors are unobserved, we can fix their location and scale arbitrarily, so we assume they are in standardized form with mean zero and standard deviation one. (We will also assume they are uncorrelated although this is not an essential requirement.)

- We assume that the residual terms are uncorrelated with each other and with the common factors. This implies that, given the values of the factors, the manifest variables are independent so that the correlations of the observed variables arise from their relationships with the factors.
- Since the factors are unobserved, the factor loadings cannot be estimated in the same way as are regression coefficients in multiple regression.
- But with the assumptions above, the factor model implies that the population covariance matrix of the observed variables, Σ, has the form

$$\Sigma = \Lambda\Lambda^{\mathrm{T}} + \Psi$$

where Ψ is a diagonal matrix containing the variances of the disturbance terms on its main diagonal (the *specific variances*) and this relationship *can* be used as the basis for estimating both the factor loadings and the specific variances.
- There are several approaches to estimation, of which the most popular are *principal factor analysis* and *maximum likelihood factor analysis*. Details of both are given in Everitt and Dunn (2001), but in essence the first operates much like principal component analysis but only tries to account for the common factor variance, and the second relies on assuming that the observed variables have a multivariate normal distribution. Maximum likelihood factor analysis includes a formal testing procedure for the number of common factors.
- The initial factor solution obtained from either method may often be simplified as an aid to interpretation by the process known as *factor rotation*. Rotation does not alter the overall structure of a solution, but only how the solution is described; rotation of factors is a process by which a solution is made more interpretable without changing its underlying mathematical properties. Rotated factors may be constrained to be independent (*orthogonal*) or allowed to correlate (*oblique*) although in practice both will often lead to the same conclusions. For a discussion of rotation, see Preacher and MacCallum (2003).
- In most applications, factor analysis stops after the estimation of the parameters in the model, the rotation of the factors, and the attempted interpretation of the fitted model. In some applications, however, the researcher might be interested in finding

the "scores" of each sample member on the common factors. But calculation of factor scores is far more problematical than that for principal component scores and will not be considered in this chapter. Some discussion of the possibilities can be found in Everitt and Dunn (2001).

11.2.3 Factor Analysis and Principal Component Compared

Factor analysis, like principal component analysis, is an attempt to explain a set of data in terms of a smaller number of dimensions than one begins with, but the procedures used to achieve this goal are essentially quite different in the two methods. Factor analysis, unlike principal component analysis, begins with a hypothesis about the covariance (or correlational) structure of the variables. The hypothesis is that a set of k latent variables exists ($k < q$), and these are adequate to account for the interrelationships of the variables although not for their full variances. Principal component analysis, however, is merely a transformation of the data, and no assumptions are made about the form of the covariance matrix from which the data arise. This type of analysis has no part corresponding to the specific variates of factor analysis. Consequently, if the factor model holds but the variances of the specific variables are small, we would expect both forms of analysis to give similar results. If, however, the specific variances are large, they will be absorbed into all the principal components, both retained and rejected, whereas factor analysis makes special provision for them.

It should be remembered that both forms of analysis are similar in one important respect, namely, that they are both pointless if the observed variables are almost uncorrelated — factor analysis because it has nothing to explain and principal components analysis because it would simply lead to components that are similar to the original variables.

11.3 Analysis Using SPSS

11.3.1 Crime in the U.S.

The **Data View** spreadsheet for the crime data in Table 11.1 contains seven variables relating to crime rates for different crime categories (**murder, rape, robbery, assault, burglary, larceny, m theft**) and a string state identifier (**state**).

The main aim of our analysis will be to identify patterns of crime in different states, but we start by generating simple descriptive summaries for each of the crime variables (shown in Display 11.1). These statistics

Descriptive Statistics

	N	Minimum	Maximum	Mean	Std. Deviation
Murder	51	1.0	31.0	7.251	4.8169
Rape	51	11.6	72.7	34.218	14.5709
Robbery	51	7	754	154.10	137.816
Aggrevated assault	51	32	668	283.35	148.339
Burglary	51	385	2221	1207.80	421.073
Larceny or theft	51	1358	4372	2941.94	763.384
Motor vehicle theft	51	99	975	393.84	223.623
Valid N (listwise)	51				

Correlation Matrix

		Murder	Rape	Robbery	Aggrevated assault	Burglary	Larceny or theft	Motor vehicle theft
Correlation	Murder	1.000	.578	.804	.781	.580	.361	.573
	Rape	.578	1.000	.530	.659	.721	.635	.569
	Robbery	.804	.530	1.000	.740	.551	.399	.786
	Aggrevated assault	.781	.659	.740	1.000	.711	.512	.638
	Burglary	.580	.721	.551	.711	1.000	.765	.578
	Larceny or theft	.361	.635	.399	.512	.765	1.000	.386
	Motor vehicle theft	.573	.569	.786	.638	.578	.386	1.000

Display 11.1   Descriptive statistics for the crime rate variables.

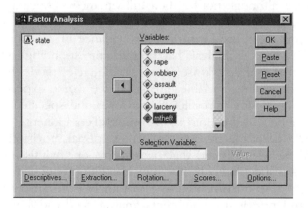

Display 11.2   Declaring the manifest variables of a factor model.

summarize the univariate distributions of the rates for each of the crime categories. A correlation matrix of the data (also included in Display 11.1) shows that the correlations between the crime rates for different types of crime are substantial suggesting that some simplification of the data using a principal component analysis *will* be possible.

In SPSS principal component analysis is classed as a form of factor analysis and is requested by using the commands

Analyze – Data Reduction – Factor...

The resulting dialogue box is shown in Display 11.2. This is used to declare the variables to be factor-analyzed (or to declare the variables that

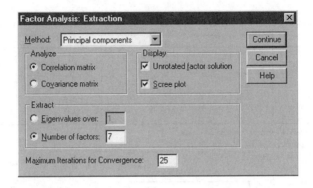

Display 11.3 **Declaring a principal component analysis and deciding the number of components.**

are to be transformed into principal components). A principal component analysis can then be specified by checking the **Extraction...** button and opting for **Principal components** in the **Method** list of the **Extraction** sub-dialogue box (Display 11.3).

Here we opt to analyze the **Correlation matrix** since the variances of rates for different types of crime differ considerably (e.g., murders are relatively rare and show small variability, while larceny/theft rates are relatively high and variable, see Display 11.1). Working with the correlation matrix amounts to using the crime rates after standardizing each to have unit standard deviation (Display 11.3). This seems sensible since without standardization the derived components are likely to be dominated by single variables with large variances.

To begin, we ask for all (seven) components to be extracted (set **Number of factors** to 7), but we also check **Screen plot** to use as a guide as to whether fewer than seven components might give an adequate representation of the data. (Note that SPSS also provides the opportunity for choosing the number of principal components according to the number of eigenvalues above a threshold, e.g., above one, which in the case of the correlation matrix amounts to identifying components with variances more than the average.)

Finally, we request the **Unrotated factor solution** since this provides the numerical values of the coefficients that define the components and so may help us to interpret them in some way. The resulting output is shown in Displays 11.4 and 11.5.

The principal components output in Display 11.4 starts with a "Component Matrix" table. The coefficients in this table specify the linear function of the observed variables that define each component. SPSS presents the coefficients scaled so that when the principal component analysis is based on the correlation matrix, they give the correlations between the observed variables and the principal components (see Box 11.1). These coefficients

Component Matrix[a]

	Component						
	1	2	3	4	5	6	7
Murder	.825	-.347	-.367	1.867E-02	.135	-.156	.147
Rape	.817	.276	-1.18E-02	-.487	.122	2.960E-02	-4.63E-02
Robbery	.846	-.417	8.929E-02	.161	.189	3.248E-02	-.200
Aggravated assault	.887	-.123	-.228	1.857E-02	-.277	.261	2.483E-03
Burglary	.854	.365	-8.62E-03	9.512E-02	-.226	-.263	-8.68E-02
Larceny or theft	.695	.623	5.302E-02	.263	.189	.120	7.865E-02
Motor vehicle theft	.793	-.281	.516	-4.29E-02	-7.98E-02	-1.68E-02	.130

Extraction Method: Principal Component Analysis.
a. 7 components extracted.

Communalities

	Initial
Murder	1.000
Rape	1.000
Robbery	1.000
Aggravated assault	1.000
Burglary	1.000
Larceny or theft	1.000
Motor vehicle theft	1.000

Extraction Method: Principal Component Analysis.

Total Variance Explained

Component	Initial Eigenvalues			Extraction Sums of Squared Loadings		
	Total	% of Variance	Cumulative %	Total	% of Variance	Cumulative %
1	4.694	67.056	67.056	4.694	67.056	67.056
2	.986	14.091	81.147	.986	14.091	81.147
3	.464	6.633	87.780	.464	6.633	87.780
4	.344	4.917	92.697	.344	4.917	92.697
5	.239	3.413	96.110	.239	3.413	96.110
6	.178	2.544	98.653	.178	2.544	98.653
7	9.426E-02	1.347	100.000	9.426E-02	1.347	100.000

Extraction Method: Principal Component Analysis.

Display 11.4   Principal component analysis output for crime rate variables.

are often used to interpret the principal components and possibly give them names, although such *reification* is not recommended and the following quotation from Marriott (1974) should act as a salutary warning about the dangers of over-interpretation of principal components:

It must be emphasised that no mathematical method is, or could be, designed to give physically meaningful results. If a mathematical expression of this sort has an obvious physical meaning, it must be attributed to a lucky chance, or to the fact that the data have a strongly marked structure that shows up in analysis. Even in the latter case, quite small sampling fluctuations can upset the interpretation, for example, the first two principal components may appear in reverse order, or may become confused altogether. Reification then requires considerable skill

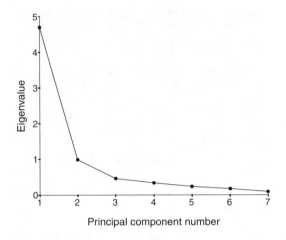

Display 11.5 Scree plot for crime rate variables.

and experience if it is to give a true picture of the physical meaning of the data.

The table labeled "Communalities" is of more interest when using exploratory factor analysis and so will be explained later in the chapter.

The final table in Display 11.4, labeled "Total Variance Explained," shows how much of the total variance of the observed variables is explained by each of the principal components. The first principal component (scaled eigenvector), by definition the one that explains the largest part of the total variance, has a variance (eigenvalue) of 4.7; this amounts to 67% of the total variance. The second principal component has a variance of about one and accounts for a further 14% of the variance and so on. The "Cumulative %" column of the table tells us how much of the total variance can be accounted for by the first k components together. For example, the first two (three) principal components account for 81% (88%) of the total variance.

The scree plot (Display 11.5) demonstrates this distribution of variance among the components graphically. For each principal component, the corresponding eigenvalue is plotted on the y-axis. By definition the variance of each component is less than the preceeding one, but what we are interested in is the "shape" of the decrease. If the curve shows an "elbow" at a given value on the x-axis, this is often taken as indicating that higher order principal components contribute a decreasing amount of additional variance and so might not be needed. Here there appears to be a marked decrease in downward slope after the second or third principal component implying that we can summarize our seven crime variables by the first two or three principal components. To simplify matters we shall assume that the two-component solution is adequate.

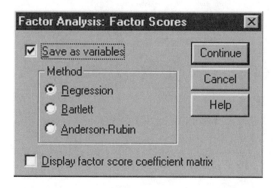

Display 11.6 Saving factor scores.

Having decided on the two-component solution, we can return to the "Component Matrix" table to try to interpret both components. The first has a high positive correlation with each of the crime variables and is simply a weighted average of the crime rates and so provides a measure of a state's overall level of crime. Such a size component always occurs when the elements of the correlation matrix are all positive as here (see Display 11.1). The second principal component is positively correlated with rape, burglary, and larceny/theft rates and negatively correlated with murder, robbery, aggravated assault, and motor vehicle theft rates. It is tempting to conclude that it reflects a differential between petty and capital crimes; sadly, the loadings on rape and car theft show that things are not so simple and the example illustrates the potential difficulties of trying to label components.

To use the first two principal component scores to display the crime rate data in a scatterplot, we first have to include them in the **Data View** spreadsheet by clicking the **Scores...** button on the **Factor Analysis** dialogue box and checking **Save as variables** in the resulting **Factor Scores** sub-dialogue box (Display 11.6). (For computing component scores, we can use the default settings; the options available apply only to the computation of factor scores.) The new spreadsheet variables are labelled fac1_1 to fac7_1 and are automatically scaled to unit standard deviation.

We can now plot the data in the space of the first two components using the following steps:

■ In the **Simple Scatterplot** dialogue box set **Y Axis** to fac1_1, **X Axis** to fac2_1 and **Label Cases by** to state (see Chapter 2, Display 2.20). This generates an initial scattergraph.

■ Open the **Chart Editor**, use the commands **Chart – Options...** from the menu bar, and set **Case Labels** to **On** and **Source of Labels** to **ID variable** in the resulting dialogue box to turn the markers on.

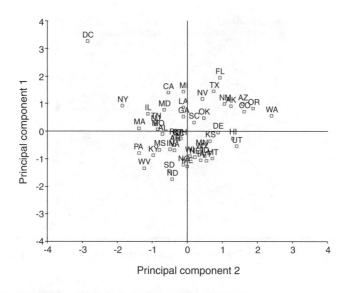

Display 11.7   Scatterplot of the first two principal components.

■ Use the **Chart Editor** to insert some reference lines that cross the *x*-
   and *y*-axes at their origins (using the commands **Chart – Reference
   Line... – Y Scale** (or **X Scale**) **– Position of Line(s): = 0**).

■ The graph is then edited a little to improve its appearance.

The principal components plot given in Display 11.7 serves to help
visualize the crime patterns of the fifty states. Scores on the *y*-axis indicate
the overall level of crime, while scores on the *x*-axis indicate (perhaps) the
differential between petty and capital crime rates. The District of Columbia
(DC) clearly stands out from the other U.S. states with the highest overall
crime rates and a large differential between petty and capital crimes in favor
of capital crimes (highest rates of murder, robbery, aggravated assault, and
motor vehicle theft). Also notable are New York (NY) and Washington (WA).
New York has an above average overall rate of crime together with a large
differential in rates between crime categories in favor of capital crimes
(robbery, aggravated assault, and motor vehicle theft). By contrast, Washington has a similar overall level of crime together with a large differential
in rates in favor of petty crimes (burglary, larceny, or theft).

11.3.2 AIDS Patients' Evaluations of Their Clinicians

The **Data View** spreadsheet for the survey data in Table 11.2 contains 14
variables (q1 to q14) as described in the introduction to this chapter. For

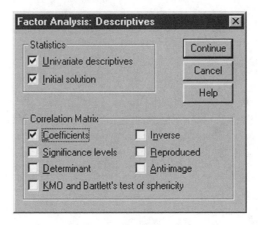

Display 11.8 **Requesting descriptive statistics via the** Factor Analysis **dialogue box.**

these data, we shall use exploratory factor analysis to investigate the underlying structure of the associations between the 14 questionnaire items. Largely as an exercise in illustrating more of the factor analysis options in SPSS, we shall use the formal test for the number of factors provided by the maximum likelihood approach (see Box 11.2) to estimate the number of factors, but then use principal factor analysis to estimate the loadings and specific variances corresponding to this number of factors.

As for principal component analysis, the 14 manifest variables to be factor-analyzed can be declared in the Factor Analysis box (see Display 11.2). Simple descriptive summaries for each of the items can be requested via the Factor Analysis box by using the Descriptives... button (Display 11.8). Here we request univariate descriptive summaries and the observed correlation matrix. (Checking Initial solution affects the factor analysis output; we will discuss this later.)

The resulting descriptive output is shown in Display 11.9. Each item has been rated on the same Likert scale (1 to 5) and the standard deviations of the item ratings did not vary a lot. It seems therefore reasonable, on this occasion, to model the covariance matrix. We also note that the correlations between the item ratings are all positive.

First, we will consider maximum likelihood (ML) factor analysis. This is a method that assumes multivariate normality and then estimates the factor model parameters by maximizing the likelihood function. We leave it as an exercise for the reader to undertake the full factor analysis with this method (see Exercise 11.4.2) and here concentrate only on its goodness-of-fit test, which assesses the null hypothesis that a model with a given number of factors explains the covariance or correlation structure of the manifest variables. The test is generally applied by successively

Descriptive Statistics

	Mean	Std. Deviation	Analysis N
friendly manner	1.66	.781	64
doubts about ability	1.92	.997	64
cold and impersonal	1.77	.868	64
reassurance	2.09	1.019	64
examines me carefully	1.94	.852	64
not enough respect	2.02	.934	64
doubt about the treatment	2.27	1.043	64
competent and well trained	1.67	.757	64
interest in me	1.91	.955	64
unanswered questions	2.27	1.144	64
jargon	2.48	1.141	64
confidence	1.94	.941	64
confide	1.88	.864	64
not free to ask questions	1.81	.941	64

Correlation Matrix

	friendly manner	doubts about ability	cold and impersonal	reassurance	examines me carefully	not enough respect	doubt about the treatment	competent and well trained	interest in me	unanswered questions	jargon	confidence	confide	not free to ask questions
friendly manner	1.000	.556	.628	.639	.516	.704	.445	.611	.786	.566	.315	.554	.688	.624
doubts about ability	.556	1.000	.584	.460	.443	.513	.478	.681	.576	.631	.271	.723	.505	.424
cold and impersonal	.628	.584	1.000	.348	.495	.494	.280	.437	.662	.399	.325	.507	.595	.334
reassurance	.639	.460	.348	1.000	.519	.515	.335	.431	.547	.550	.206	.486	.536	.466
examines me carefully	.516	.443	.495	.519	1.000	.540	.376	.558	.656	.538	.130	.629	.507	.381
not enough respect	.704	.513	.494	.515	.540	1.000	.631	.636	.642	.575	.261	.615	.730	.581
doubt about the treatment	.445	.478	.280	.335	.376	.631	1.000	.494	.344	.645	.224	.470	.443	.505
competent and well trained	.611	.681	.437	.431	.558	.636	.494	1.000	.703	.615	.242	.751	.495	.492
interest in me	.786	.576	.662	.547	.656	.642	.344	.703	1.000	.619	.174	.700	.659	.528
unanswered questions	.566	.631	.399	.550	.538	.575	.645	.615	.619	1.000	.253	.665	.532	.563
jargon	.315	.271	.325	.206	.130	.261	.224	.242	.174	.253	1.000	.310	.240	.234
confidence	.554	.723	.507	.486	.629	.615	.470	.751	.700	.665	.310	1.000	.654	.507
confide	.688	.505	.595	.536	.507	.730	.443	.495	.659	.532	.240	.654	1.000	.557
not free to ask questions	.624	.424	.334	.466	.381	.581	.505	.492	.528	.563	.234	.507	.557	1.000

Display 11.9   Descriptive statistics for questionnaire variables.

increasing the number of factors by one until the test yields an insignificant result. However, like all tests for model assumptions, the procedure is not without difficulties. For small samples, too few factors might be indicated and for large samples, too many. In addition, it is not clear what single test level should be chosen and the test has been shown very sensitive to departures from normality. Here we demonstrate the test procedure in the spirit of an explorative tool rather than a formal test.

Parameter estimation via maximum likelihood is specified via the **Extraction** sub-dialogue box (see Display 11.3). As soon as **ML** is selected from the **Method** list, it is no longer possible to choose the covariance matrix since SPSS automatically models the correlation matrix. Although our later analysis will concentrate on the former, given the small differences in standard deviations for the questionnaire variables, this should not present a real problem. By fixing the **Number of factors**, we can fit a series of factor model with one, two, and three factors.

The resulting series of tests for model fit is shown in Display 11.10. We detect significant lack of model fit for the one- and two-factor solutions with the three-factor model failing to detect departure from the model at the 5% level ($X^2(1) = 69$, $p = 0.058$). So, according to the ML test criterion, a three-factor model seems reasonable for these data.

a) One-factor model

Goodness-of-fit Test

Chi-Square	df	Sig.
139.683	77	.000

b) Two-factor model

Goodness-of-fit Test

Chi-Square	df	Sig.
101.589	64	.002

a) Three-factor model

Goodness-of-fit Test

Chi-Square	df	Sig.
68.960	52	.058

Display 11.10   ML-based tests of model fit obtained for a series of factor models for questionnaire variables.

We now employ principal factor analysis as our method for *factor extraction*, i.e., fitting the factor analysis model by estimating factor loadings and specific variances. Principal factor analysis is similar in many respects to principal component analysis but uses what is sometimes known as the *reduced* covariance or correlation matrix as the basis of the calculations involved, i.e., the sample covariance or correlation matrix with estimates of *communalities* on the main diagonal, where the communality of a variable is the variance accountable for by the common factors. One frequently used estimate of a variable's communality is the square of the multiple correlation of the variable with the other observed variables. This is the method used in SPSS. From an initial version of the reduced correlation or covariance matrix, factor loadings can be found and these can then be used to update the communalities and so on until some convergence criterion is satisfied (full details are given in Everitt and Dunn, 2001).

The extraction method can be specified by the **Method** setting of the **Extraction** sub-dialogue box (see Display 11.3); SPSS refers to principal factor analysis as *principal axis factoring*. For principal factor analysis, there is a choice between basing the analysis on the correlation or the covariance matrix and we choose **Covariance matrix.** We also set the **Number of factors** to **3** to fit a three-factor model and, finally, check **Unrotated factor solution** so that the unrotated factor solution output will be given.

The resulting principal factor analysis output is shown in Display 11.11. The "Communalities" table shows the communality estimates before and

Communalities

	Raw		Rescaled	
	Initial	Extraction	Initial	Extraction
friendly manner	.296	.511	.485	.837
doubts about ability	.686	.670	.690	.674
cold and impersonal	.371	.391	.492	.519
reassurance	.581	.455	.559	.438
examines me carefully	.298	.351	.410	.483
not enough respect	.552	.624	.633	.715
doubt about the treatme	.711	.806	.654	.742
competent and well trained	.237	.385	.413	.671
interest in me	.681	.736	.747	.808
unanswered questions	1.168	.896	.892	.684
jargon	.429	.139	.330	.107
confidence	.614	.696	.694	.786
confide	.392	.495	.525	.664
not free toask questions	.396	.454	.447	.513

Extraction Method: Principal Axis Factoring.

Display 11.11   Principal factor analysis output for questionnaire variables.

after factor extraction and on the original ("Raw") variance scale and after standardization to unit standard deviation ("Rescaled"). From this table, we see, for example, that little of the variance of the item "jargon" (only 10.7%) can be attributed to the three common factors. (This is a convenient point to note that the "Communalities" table given when using principal components: Display 11.4 simply contains all ones since we requested all seven components and these will explain the total variance of the observed variables.)

The "Total Variance Explained" table lists the variances of the principal components and those of the extracted factors. The percentages of total variance accounted for by the principal components are included since they provide an easy-to-compute criteria for choosing the number of factors (and not because the principal components themselves are of interest). According to the principal component analysis part of the table our choice of three factors seems reasonable — the first three eigenvalues are each greater than the average variance (0.9). Note that the percentage of total variance accounted for by the extracted factors tends to be lower than that of the same number of principal components since factor analysis deals with common factor variance rather than total variance.

Finally, the "Factor Matrix" table gives the estimated factor loadings of our three-factor model for the covariance matrix of the questionnaire variables. When the covariance matrix is modeled, the factor loadings correspond to the (estimated) covariances between the observed variables

Total Variance Explained

	Factor	Initial Eigenvalues[a]			Extraction Sums of Squared Loadings		
		Total	% of Variance	Cumulative %	Total	% of Variance	Cumulative %
Raw	1	6.822	53.741	53.741	6.499	51.198	51.198
	2	1.190	9.372	63.113	.674	5.310	56.508
	3	.969	7.637	70.750	.437	3.440	59.948
	4	.741	5.834	76.584			
	5	.629	4.954	81.538			
	6	.458	3.608	85.146			
	7	.435	3.425	88.571			
	8	.381	2.998	91.569			
	9	.304	2.395	93.964			
	10	.286	2.252	96.216			
	11	.171	1.345	97.562			
	12	.122	.960	98.521			
	13	.106	.836	99.357			
	14	8.160E-02	.643	100.000			
Rescaled	1	6.822	53.741	53.741	7.412	52.945	52.945
	2	1.190	9.372	63.113	.699	4.995	57.941
	3	.969	7.637	70.750	.531	3.790	61.731
	4	.741	5.834	76.584			
	5	.629	4.954	81.538			
	6	.458	3.608	85.146			
	7	.435	3.425	88.571			
	8	.381	2.998	91.569			
	9	.304	2.395	93.964			
	10	.286	2.252	96.216			
	11	.171	1.345	97.562			
	12	.122	.960	98.521			
	13	.106	.836	99.357			
	14	8.160E-02	.643	100.000			

Extraction Method: Principal Axis Factoring.

[a.] When analyzing a covariance matrix, the initial eigenvalues are the same across the raw and rescaled solution.

Factor Matrix[a]

	Raw			Rescaled		
	Factor			Factor		
	1	2	3	1	2	3
friendly manner	.650	.170	.244	.832	.218	.313
doubts about ability	.749	.001	-.329	.752	.001	-.330
cold and impersonal	.561	.276	-.030	.646	.317	-.035
reassurance	.658	.076	.126	.646	.075	.124
examines me carefully	.578	.124	-.035	.679	.146	-.041
not enough respect	.754	-.081	.223	.807	-.086	.238
doubt about the treatment	.682	-.581	.063	.654	-.557	.060
competent and well trained	.594	.002	-.178	.785	.003	-.235
interest in me	.792	.329	.018	.830	.345	.019
unanswered questions	.906	-.252	-.110	.791	-.220	-.096
jargon	.372	-.022	-.024	.326	-.020	-.021
confidence	.784	.054	-.279	.834	.057	-.297
confide	.668	.117	.189	.773	.136	.218
not free to ask questions	.632	-.123	.197	.672	-.131	.210

Extraction Method: Principal Axis Factoring.

[a.] 3 factors extracted .23 iterations required.

Display 11.11 (continued)

and the factors, and these are given in the "Raw" part of the table. To be able to compare factor loadings across variables, we need to standardize the effects to unit standard deviations of the observed variables. In other words, we need to transform the raw factor loadings into correlations. These are given in the "Rescaled" part of the table. The solution of a factor

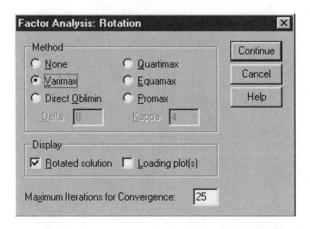

Display 11.12   Requesting a varimax rotation of the initial factor solution.

analysis is not unique and we delay naming the factors until after factor rotation.

When applying exploratory factor analysis, an attempt is always made to describe the solution as simply as possible by applying some method of factor rotation (see Box 11.2). Interpretation is more straightforward if each variable is highly loaded on at most one factor, and if all factor loadings are either large and positive or near zero, with few intermediate values (see Everitt and Dunn, 2001). SPSS provides several methods of rotation that try to achieve these goals, some of which produce orthogonal factors (*varimax, quartimax,* and *equamax*) and others that lead to an oblique solution (*direct oblimin* and *promax*). Here we shall employ the varimax procedure. (For technical details of how varimax rotation sets out to achieve its aim of simplifying the initial factor solution, see Mardia, Kent, and Bibby, 1979; for details of other rotation techniques, see Preacher and MacCullum, 2003.)

The **Rotation...** button on the **Factor Analysis** dialogue box opens the **Rotation** sub-dialogue box which allows specification of the rotation method (see Display 11.12). (We choose not to use a *loadings plot*; this is a graphical display of the rotated solution, although, in our view not particularly helpful for interpretation, especially for more than two dimensions.)

The "Rotated Factor" matrix is shown in Display 11.13. It is this matrix that is examined with a view to interpreting the factor analysis. Such interpretation may be further aided picking out "high" loadings. In practice, a largely arbitrary threshold value of 0.4 is often equated to "high" in this context. In SPSS "unimportant" loadings can be suppressed and factor loadings reordered according to size, as shown in Display 11.14.

Rotated Factor Matrix[a]

	Raw			Rescaled		
	Factor			Factor		
	1	2	3	1	2	3
friendly manner	.635	.222	.240	.813	.285	.308
doubts about ability	.276	.704	.313	.276	.706	.314
cold and impersonal	.476	.403	.050	.548	.464	.057
reassurance	.523	.305	.298	.513	.299	.293
examines me carefully	.404	.392	.184	.474	.460	.216
not enough respect	.556	.259	.498	.595	.277	.533
doubt about the treatment	.156	.258	.846	.150	.247	.811
competent and well trained	.267	.494	.264	.353	.652	.349
interest in me	.676	.512	.133	.708	.536	.140
unanswered questions	.367	.582	.650	.321	.509	.568
jargon	.206	.236	.203	.180	.207	.178
confidence	.353	.694	.298	.376	.738	.316
confide	.587	.268	.282	.679	.310	.327
not free to ask questions	.443	.200	.466	.471	.212	.496

Extraction Method: Principal Axis Factoring.
Rotation Method: Varimax with Kaiser Normalization.
a. Rotation converged in 9 iterations.

Display 11.13   Rotated factor analysis solution for questionnaire variables.

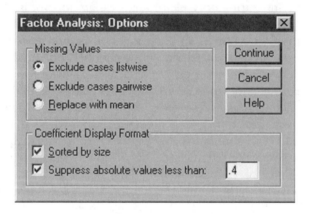

Display 11.14   Suppressing unimportant factor loadings.

The "edited" rotated solution is shown in Display 11.15. We use the rescaled factor loadings (correlations) to evaluate which variables load on each factor and so to interpret the meaning of the factor. The rotated factors might perhaps be labeled "evaluation of doctor's interpersonal skills," "confidence in doctor's ability" and "confidence in recommended treatment." We note that, even after suppressing low loadings, some

Rotated Factor Matrix[a]

	Raw Factor			Rescaled Factor		
	1	2	3	1	2	3
friendly manner	.635			.813		
interest in me	.676	.512		.708	.536	
confide	.587			.679		
not enough respect	.556		.498	.595		.533
cold and impersonal	.476	.403		.548	.464	
reassurance	.523			.513		
examines me carefully	.404	.392		.474	.460	
confidence		.694			.738	
doubts about ability		.704			.706	
competent and well trained		.494			.652	
jargon						
doubt about the treatment			.846			.811
unanswered questions		.582	.650		.509	.568
not free to ask questions	.443		.466	.471		.496

Extraction Method: Principal Axis Factoring.
Rotation Method: Varimax with Kaiser Normalization.

[a.] Rotation converged in 9 iterations.

Display 11.15   Edited rotated factor analysis solution for questionnaire variables.

variables load on several factors. For example, careful examination appears to be affected by factors 1 and 2. We also find that question 11, the use of jargon by the clinician, does not seem to probe any of these concepts and should, perhaps, be excluded from the questionnaire.

The goodness of fit of a factor analysis model can (and should be) examined by comparing the observed correlation or covariance matrix with that predicted by the model, i.e., the *reproduced correlation* or *covariance matrix.* The latter can be requested in SPSS by checking **Reproduced** in the **Descriptives** sub-dialogue box (see Display 11.8). The covariance matrix predicted by our fitted factor analysis model is shown in Display 11.16. Also included is the *residual covariance matrix,* the entries of which are simply the differences between the observed and predicted covariances. If the factor analysis model fits well, the elements of the residual correlation matrix will be small. Here, careful examination of the residual matrix shows that all deviations are less than 0.08 except for the covariance between use of jargon and being cold and impersonal, which is underestimated by 0.12. Given our previous comments on the jargon variable and an average variance of 0.9, these values are probably sufficiently small to claim that the three-factor model gives an adequate description of the data.

Reproduced Covariances

	friendly manner	doubts about ability	cold and impersonal	reassurance	examines me carefully	not enough respect	doubt about treatment	competent and well trained	interest in me	unanswered questions	jargon	confidence	confide	not free to ask question
Reproduced Cov friendly manner	.511b	.407	.404	.472	.388	.531	.360	.343	.575	.519	.232	.450	.500	.438
doubts about abil	.407	.670b	.430	.452	.445	.491	.489	.504	.588	.714	.286	.679	.439	.409
cold and imperso	.404	.430	.391b	.386	.360	.394	.220	.339	.534	.442	.203	.463	.401	.315
reassurance	.472	.452	.386	.455b	.386	.518	.412	.369	.549	.563	.240	.485	.473	.432
examines me car	.388	.445	.360	.386	.351b	.418	.320	.350	.498	.496	.213	.470	.394	.343
not enough respe	.531	.491	.394	.518	.418	.624b	.575	.408	.575	.678	.277	.525	.536	.531
doubt about the tr	.360	.489	.220	.412	.320	.575	.806b	.393	.350	.757	.265	.486	.399	.515
competent and w trained	.343	.504	.339	.369	.350	.408	.393	.385b	.468	.557	.225	.516	.364	.340
interest in me	.575	.588	.534	.549	.498	.575	.350	.468	.736b	.633	.287	.634	.571	.464
unanswered ques	.519	.714	.442	.563	.496	.678	.757	.557	.633	.896b	.345	.727	.555	.582
jargon	.232	.286	.203	.240	.213	.277	.265	.225	.287	.345	.139b	.297	.241	.233
confidence	.450	.679	.463	.485	.470	.525	.486	.516	.634	.727	.297	.696b	.477	.434
confide	.500	.439	.401	.473	.394	.536	.399	.364	.571	.555	.241	.477	.495b	.445
not free to ask qu	.438	.409	.315	.432	.343	.531	.515	.340	.464	.582	.233	.434	.445	.454
Residuala friendly manner		620E-02	220E-02	726E-02	504E-02	718E-02	3.009E-03	860E-02	.112E-02	.334E-02	32E-02	74E-02	58E-02	.022E-02
doubts about abil	20E-02		486E-02	567E-02	903E-02	400E-02	7.958E-03	784E-03	.011E-02	.153E-03	80E-02	99E-03	1E-03	.085E-02
cold and imperso	20E-02	486E-02		843E-02	566E-03	820E-03	3.346E-02	220E-02	.474E-02	.506E-02	.119	12E-02	33E-02	.180E-02
reassurance	26E-02	567E-02	843E-02		459E-02	770E-02	5.679E-02	618E-02	.635E-02	7.815E-02	06E-04	01E-02	7E-04	.467E-02
examines me car	50E-02	903E-02	566E-03	459E-02		157E-02	.456E-02	951E-03	.533E-02	.860E-02	71E-02	03E-02	1E-02	.776E-02
not enough respe	72E-02	400E-02	820E-03	770E-02	157E-02		4.007E-02	164E-02	.728E-03	.361E-02	21E-03	13E-02	7E-02	.959E-02
doubt about the tr	09E-03	958E-03	346E-02	679E-02	456E-02	007E-02		389E-03	.270E-03	.291E-02	00E-03	55E-02	42E-04	.993E-02
competent and w trained	60E-02	784E-03	220E-02	618E-02	951E-03	164E-02	2.389E-03		.006E-02	.413E-02	61E-02	85E-02	02E-02	.974E-02
interest in me	12E-02	011E-02	474E-02	635E-02	533E-02	728E-02	7.270E-03	006E-02		4.348E-02	78E-02	69E-02	74E-02	.033E-02
unanswered ques	33E-02	153E-03	506E-02	815E-02	860E-02	361E-02	.291E-02	413E-02	.348E-02		54E-02	04E-02	89E-02	.430E-02
jargon	32E-02	180E-02	.119	065E-04	707E-02	321E-03	.200E-03	610E-02	.781E-02	.536E-02		25E-02	09E-03	.793E-02
confidence	37E-02	899E-03	912E-02	901E-02	403E-02	613E-02	2.455E-02	885E-02	.969E-03	.204E-02	25E-02		36E-02	.436E-02
confide	58E-02	111E-03	533E-02	167E-04	112E-02	317E-02	4.420E-04	016E-02	.745E-02	.889E-02	09E-02	36E-02		.258E-03
not free to ask qu	22E-02	085E-02	180E-02	467E-02	776E-02	959E-02	.993E-02	974E-03	.033E-02	.430E-02	93E-02	36E-02	58E-03	

Extraction Method: Principal Axis Factoring.
a. Residuals are computed between observed and reproduced covariances.
b. Reproduced communalities

Display 11.16   Reproduced covariance matrix and residual covariance matrix for questionnaire variables.

11.4 Exercises

11.4.1 Air Pollution in the U.S.

The data below relate to air pollution in 41 U.S. cities. Run a principal component analysis on variables $X_1, \ldots, X_6$ and save the first two principal component scores for each city. Find the correlation between each principal component and the level of air pollution (Y). Use the scores to produce a scatterplot of the cities and interpret your results particularly in terms of the air pollution level in the cities.

The variables are as follows:

Y: SO$_2$ content of air in micrograms per cubic meter
X_1: Average annual temperature in degrees F
X_2: Number of manufacturing enterprises employing 20 or more workers
X_3: Population size in thousands
X_4: Average annual wind speed in miles per hour
X_5: Average annual precipitation in inches
X_6: Average number of days with precipitation per year

Table 11.3 Air Pollution Data

	Y	X_1	X_2	X_3	X_4	X_5	X_6
Phoenix	10	70.3	213	582	6.0	7.05	36
Little Rock	13	61.0	91	132	8.2	48.52	100
San Francisco	12	56.7	453	716	8.7	20.66	67
Denver	17	51.9	454	515	9.0	12.95	86
Hartford	56	49.1	412	158	9.0	43.37	127
Wilmington	36	54.0	80	80	9.0	40.25	114
Washington	29	57.3	434	757	9.3	38.89	111
Jacksonville	14	68.4	136	529	8.8	54.47	116
Miami	10	75.5	207	335	9.0	59.80	128
Atlanta	24	61.5	368	497	9.1	48.34	115
Chicago	110	50.6	3344	3369	10.4	34.44	122
Indianapolis	28	52.3	361	746	9.7	38.74	121
Des Moines	17	49.0	104	201	11.2	30.85	103
Wichita	8	56.6	125	277	12.7	30.58	82
Louisville	30	55.6	291	593	8.3	43.11	123
New Orleans	9	68.3	204	361	8.4	56.77	113
Baltimore	47	55.0	625	905	9.6	41.31	111
Detroit	35	49.9	1064	1513	10.1	30.96	129
Minneapolis	29	43.5	699	744	10.6	25.94	137
Kansas City	14	54.5	381	507	10.0	37.00	99
St. Louis	56	55.9	775	622	9.5	35.89	105
Omaha	14	51.5	181	347	10.9	30.18	98
Albuquerque	11	56.8	46	244	8.9	7.77	58
Albany	46	47.6	44	116	8.8	33.36	135
Buffalo	11	47.1	391	463	12.4	36.11	166
Cincinnati	23	54.0	462	453	7.1	39.04	132
Cleveland	65	49.7	1007	751	10.9	34.99	155
Columbus	26	51.5	266	540	8.6	37.01	134
Philadelphia	69	54.6	1692	1950	9.6	39.93	115
Pittsburgh	61	50.4	347	520	9.4	36.22	147
Providence	94	50.0	343	179	10.6	42.75	125
Memphis	10	61.6	337	624	9.2	49.10	105
Nashville	18	59.4	275	448	7.9	46.00	119
Dallas	9	66.2	641	844	10.9	35.94	78
Houston	10	68.9	721	1233	10.8	48.19	103
Salt Lake City	28	51.0	137	176	8.7	15.17	89
Norfolk	31	59.3	96	308	10.6	44.68	116
Richmond	26	57.8	197	299	7.6	42.59	115
Seattle	29	51.1	379	531	9.4	38.79	164
Charleston	31	55.2	35	71	6.5	40.75	148
Milwaukee	16	45.7	569	717	11.8	29.07	123

Source: Hand et al., 1994.

11.4.2 More on AIDS Patients' Evaluations of Their Clinicians: Maximum Likelihood Factor Analysis

Use maximum likelihood factor analysis to model the correlations between the AIDS patients' ratings of their clinicians (Table 11.2). After a varimax rotation, compare your findings with those from the principal factor analysis described in the main body of the chapter.

Investigate the use of other rotation procedures (both orthogonal and oblique) on both the maximum likelihood solution and the principal factor analysis solution.

Chapter 12

Classification: Cluster Analysis and Discriminant Function Analysis; Tibetan Skulls

12.1 Description of Data

The data to be used in this chapter are shown in Table 12.1. These data, collected by Colonel L.A. Waddell, were first reported in Morant (1923) and are also given in Hand et al. (1994). The data consist of five measurements on each of 32 skulls found in the southwestern and eastern districts of Tibet. The five measurements (all in millimeters) are as follows:

- Greatest length of skull (measure 1)
- Greatest horizontal breadth of skull (measure 2)
- Height of skull (measure 3)
- Upper face length (measure 4)
- Face breadth between outermost points of cheek bones (measure 5)

According to Morant (1923), the data can be divided into two groups. The first comprises skulls 1 to 17 found in graves in Sikkim and the neighboring area of Tibet (Type A skulls). The remaining 15 skulls (Type B skulls) were picked up on a battlefield in the Lhasa district and are believed

Table 12.1   Tibetan Skulls Data

Skull Type	Measure 1	Measure 2	Measure 3	Measure 4	Measure 5
A	190.5	152.5	145	73.5	136.5
A	172.5	132	125.5	63	121
A	167	130	125.5	69.5	119.5
A	169.5	150.5	133.5	64.5	128
A	175	138.5	126	77.5	135.5
A	177.5	142.5	142.5	71.5	131
A	179.5	142.5	127.5	70.5	134.5
A	179.5	138	133.5	73.5	132.5
A	173.5	135.5	130.5	70	133.5
A	162.5	139	131	62	126
A	178.5	135	136	71	124
A	171.5	148.5	132.5	65	146.5
A	180.5	139	132	74.5	134.5
A	183	149	121.5	76.5	142
A	169.5	130	131	68	119
A	172	140	136	70.5	133.5
A	170	126.5	134.5	66	118.5
B	182.5	136	138.5	76	134
B	179.5	135	128.5	74	132
B	191	140.5	140.5	72.5	131.5
B	184.5	141.5	134.5	76.5	141.5
B	181	142	132.5	79	136.5
B	173.5	136.5	126	71.5	136.5
B	188.5	130	143	79.5	136
B	175	153	130	76.5	142
B	196	142.5	123.5	76	134
B	200	139.5	143.5	82.5	146
B	185	134.5	140	81.5	137
B	174.5	143.5	132.5	74	136.5
B	195.5	144	138.5	78.5	144
B	197	131.5	135	80.5	139
B	182.5	131	135	68.5	136

to be those of native soldiers from the eastern province of Khams. These skulls were of particular interest since it was thought at the time that Tibetans from Khams might be survivors of a particular human type, unrelated to the Mongolian and Indian types that surrounded them.

There are two questions that might be of interest for these data:

- Do the five measurements discriminate between the two assumed groups of skulls and can they be used to produce a useful rule for classifying other skulls that might become available?
- Taking the 32 skulls together, are there any natural groupings in the data and, if so, do they correspond to the groups assumed by Morant (1923)?

12.2 Classification: Discrimination and Clustering

Classification is an important component of virtually all scientific research. Statistical techniques concerned with classification are essentially of two types. The first (*cluster analysis*; see Everitt, Landau, and Leese, 2001) aim to uncover groups of observations from initially unclassified data. There are many such techniques and a brief description of those to be used in this chapter is given in Box 12.1. The second (*discriminant function analysis*; see Everitt and Dunn, 2001) works with data that is already classified into groups to derive rules for classifying new (and as yet unclassified) individuals on the basis of their observed variable values. The most well-known technique here is *Fisher's linear discriminant function analysis* and this is described in Box 12.2.

Box 12.1 Cluster Analysis

(1) *Distance and similarity measures*

- Many clustering techniques begin not with the raw data but with a matrix of inter-individual measures of distance or similarity calculated from the raw data.
- Here we shall concentrate on distance measures, of which the most common is the *Euclidean distance* given by

$$d_{ij} = \left[\sum_{l=1}^{q} \left(x_{il} - x_{jl} \right)^2 \right]^{\frac{1}{2}}$$

where d_{ij} is the Euclidean distance for two individuals i and j, each measured on q variables, x_{il}, x_{jl}, $l = 1, \ldots, q$.

■ Euclidean distances are the starting point for many clustering techniques, but care is needed if the variables are on very different scales, in which case some form of standardization will be needed (see Everitt et al., 2001).

■ For a comprehensive account of both distance and similarity measures, see Everitt et al. (2001).

(2) Agglomerative hierarchical techniques

■ These are a class of clustering techniques that proceed by a series of steps in which progressively larger groups are formed by joining together groups formed earlier in the process.

■ The initial step involves combining the two individuals who are closest (according to whatever distance measure is being used).

■ The process goes from individuals to a final stage in which all individuals are combined, with the closest two groups being combined at each stage.

■ At each stage, more and more individuals are linked together to form larger and larger clusters of increasingly dissimilar elements.

■ In most applications of these methods, the researcher will want to determine the stage at which the solution provides the best description of the structure in the data, i.e., determine the number of clusters.

■ Different methods arise from the different possibilities for defining inter-group distance. Two widely applied methods are *complete linkage* in which the distance between groups is defined as the distance between the most remote pair of individuals, one from each group, and *average linkage* in which inter-group distance is taken as the average of all inter-individual distances made up of pairs of individuals, one from each group.

■ The series of steps in this type of clustering can be conveniently summarized in a tree-like diagram known as a *dendrogram* (examples are given in the text).

(3) k-means clustering

■ This is a method of clustering that produces a partition of the data into a particular number of groups set by the investigator.

■ From an initial partition, individuals are moved into other groups if they are "closer" to its mean vector than that of their current group (Euclidean distance is generally used here). After each move, the relevant cluster mean vectors are updated.

- The procedure continues until all individuals in a cluster are closer to their own cluster mean vector than to that of any other cluster.
- Essentially the technique seeks to minimize the variability within clusters and maximize variability between clusters.
- Finding the optimal number of groups will also be an issue with this type of clustering. In practice, a *k*-means solution is usually found for a range of values of *k*, and then one of the largely *ad hoc* techniques described in Everitt et al. (2001) for indicating the correct number of groups applied.
- Different methods of cluster analysis applied to the same set of data often result in different solutions. Many methods are really only suitable when the clusters are approximately spherical in shape, details are given in Everitt et al. (2001).

Box 12.2 Fisher's Linear Discriminant Function

- A further aspect of the classification of multivariate data concerns the derivation of rules and procedures for allocating individuals/objects to one of a set of *a priori* defined groups in some optimal fashion on the basis of a set of q measurements, $x_1, x_2, \ldots, x_q$, taken on each individual or object.
- This is the province of *assignment* or *discrimination* techniques.
- The sample of observations for which the groups are known is often called the *training set*.
- Here we shall concentrate on the two group situation and on the most commonly used assignment procedure, namely, *Fisher's linear discriminant function* (Fisher, 1936).
- Fisher's suggestion was to seek a linear transformation of the variables

$$z = a_1 x_1 + a_2 x_2 + \cdots + a_p x_q$$

such that the separation between the group means on the transformed scale, $\bar{z}_1$ and $\bar{z}_2$, would be maximized relative to the within group variation on the z-scale.

■ Fisher showed that the coefficients for such a transformation are given by

$$\mathbf{a} = \mathbf{S}^{-1}(\bar{\mathbf{x}}_1 - \bar{\mathbf{x}}_2)$$

where $\mathbf{a}^{\mathrm{T}} = [a_1, a_2, \cdots, a_q]$ and $\bar{\mathbf{x}}_1$ and $\bar{\mathbf{x}}_2$ are the group mean vectors of sample means in the two groups and $\mathbf{S}$ is the pooled within-groups covariance matrix, calculated from the separate within-group covariance matrices $\mathbf{S}_1$ and $\mathbf{S}_2$ as

$$\mathbf{S} = \frac{1}{n_1 + n_2 - 2}\left[(n_1 - 1)\mathbf{S}_1 + (n_2 - 1)\mathbf{S}_2\right]$$

where n_1 and n_2 are the sample sizes of each group.
■ If an individual has a discriminant score closer to $\bar{z}_1$ than to $\bar{z}_2$, assignment is to group 1, otherwise it is to group 2.
■ The classification rule can be formalized by defining a cut-off value $\bar{z}_c$ given by

$$z_c = \frac{\bar{z}_1 + \bar{z}_2}{2}$$

■ Now, assuming that $\bar{z}_1$ is the larger of the two means, the classification rule of an individual with discriminant score z_i is
    → Assign individual to group 1 if $z_i - z_c > 0$
    → Assign individual to group 2 if $z_i - z_c \leq 0$
 (This rule assumes that the prior probabilities of being in each group are 0.5; see Hand, 1997, for how the rule changes if the prior probabilities differ from each other.)
■ The assumptions under which Fisher's method is optimal are:
 (1) The data in both groups have a normal distribution
 (2) The covariance matrices of each group are the same
■ If the first assumption is in doubt, then a logistic regression approach might be used (see Huberty, 1994).
■ If the second assumption is not thought to hold, then a *quadratic discriminant function* can be derived without the need to pool the covariance matrices of each group. In practice, this quadratic function is not usually of great use because of the number of parameters that need to be estimated to define the function.
■ Assignment techniques when there are more than two groups, for example, *canonical variates analysis*, are described in Hand (1998).

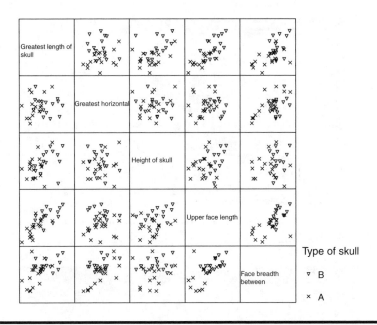

Display 12.1   Scatterplot matrix of skull measurements from Table 12.1.

12.3 Analysis Using SPSS

12.3.1 Tibetan Skulls: Deriving a Classification Rule

The **Data View** spreadsheet for the data in Table 12.1 contains the five skull measurements (**meas1** to **meas5**) and a numeric (**place**) and string version (**type**) of the grouping variable. As always, some preliminary graphical display of the data might be useful and here we will display them as a scatterplot matrix in which group membership is indicated (see Display 12.1). While this diagram only allows us to assess the group separation in two dimensions, it seems to suggest that face breadth between outermost points of cheek bones (**meas5**), greatest length of skull (**meas1**), and upper face length (**meas4**) provide the greatest discrimination between the two skull types.

We shall now use Fisher's linear discriminant function to derive a classification rule for assigning skulls to one of the two predefined groups on the basis of the five measurements available. A discriminant function analysis is requested in SPSS by the commands:

Analyze – Classify – Discriminant...

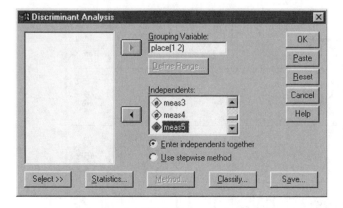

Display 12.2   Declaring a discriminant function analysis.

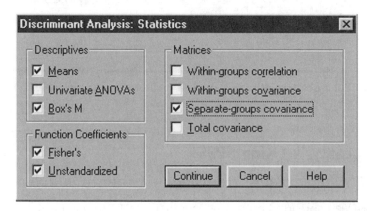

Display 12.3   Requesting descriptive statistics and Box's test of equality of co-variance matrices.

This opens the Discriminant Analysis dialogue box where the potential discriminating variables are included under the Independents list and the grouping variable under the Grouping Variable heading (Display 12.2).

The Statistics sub-dialogue box can be used to request descriptive statistics and Box's test of equality of covariance matrices (Display 12.3). We opt for Means, Separate-groups covariance and the formal test. (We also check the two options for Function Coefficients. This affects the classification output and we will discuss the related output tables later.)

The resulting descriptive output is shown in Display 12.4. Means and standard deviations of each of the five measurements for each type of skull and overall are given in the "Group Statistics" table. The within-group covariance matrices shown in the "Covariance Matrices" table suggest that

Group Statistics

place where skulls were found		Mean	Std. Deviation	Valid N (listwise) Unweighted	Weighted
Sikkem or Tibet	Greatest length of skull	174.824	6.7475	17	17.000
	Greatest horizontal breadth of skull	139.353	7.6030	17	17.000
	Height of skull	132.000	6.0078	17	17.000
	Upper face length	69.824	4.5756	17	17.000
	Face breadth between outermost points of cheek bones	130.353	8.1370	17	17.000
Lhasa	Greatest length of skull	185.733	8.6269	15	15.000
	Greatest horizontal breadth of skull	138.733	6.1117	15	15.000
	Height of skull	134.767	6.0263	15	15.000
	Upper face length	76.467	3.9118	15	15.000
	Face breadth between outermost points of cheek bones	137.500	4.2384	15	15.000
Total	Greatest length of skull	179.938	9.3651	32	32.000
	Greatest horizontal breadth of skull	139.063	6.8412	32	32.000
	Height of skull	133.297	6.0826	32	32.000
	Upper face length	72.938	5.3908	32	32.000
	Face breadth between outermost points of cheek bones	133.703	7.4443	32	32.000

Covariance Matrices

place where skulls were found		Greatest length of skull	Greatest horizontal breadth of skull	Height of skull	Upper face length	Face breadth between outermost points of cheek bones
Sikkem or Tibet	Greatest length of skull	45.529	25.222	12.391	22.154	27.972
	Greatest horizontal breadth of skull	25.222	57.805	11.875	7.519	48.055
	Height of skull	12.391	11.875	36.094	-.313	1.406
	Upper face length	22.154	7.519	-.313	20.936	16.769
	Face breadth between outermost points of cheek bones	27.972	48.055	1.406	16.769	66.211
Lhasa	Greatest length of skull	74.424	-9.523	22.737	17.794	11.125
	Greatest horizontal breadth of skull	-9.523	37.352	-11.263	.705	9.464
	Height of skull	22.737	-11.263	36.317	10.724	7.196
	Upper face length	17.794	.705	10.724	15.302	8.661
	Face breadth between outermost points of cheek bones	11.125	9.464	7.196	8.661	17.964

Display 12.4   Descriptive output for the skulls data.

the sample values differ to some extent, but according to Box's test for equality of covariances (tables "Log Determinants" and "Test Results") these differences are not statistically significant ($F(15,3490) = 1.2$, $p = 0.25$). It appears that the equality of covariance matrices assumption needed for Fisher's linear discriminant approach to be strictly correct is valid here. (In practice, Box's test is not of great use since even if it suggests a departure for the equality hypothesis, the linear discriminant may still be

Log Determinants

place where skulls were found	Rank	Log Determinant
Sikkem or Tibet	5	16.164
Lhasa	5	15.773
Pooled within-groups	5	16.727

The ranks and natural logarithms of determinants printed are those of the group covariance matric

Test Results

Box's M		22.371
F	Approx.	1.218
	df1	15
	df2	3489.901
	Sig.	.249

Tests null hypothesis of equal population covariance matric

Display 12.4 (continued)

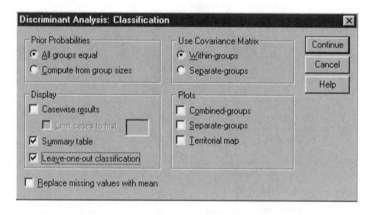

Display 12.5   Requesting a linear discriminant function.

preferable over a quadratic function.) Here we shall simply assume normality for our data relying on the robustness of Fisher's approach to deal with any minor departure from the assumption (see Hand, 1998).

The type of discriminant analysis can be declared using the **Classify...** button in the **Discriminant Analysis** dialogue box. This opens the **Classification** sub-dialogue box (Display 12.5). A linear discriminant function is constructed when the default **Within-groups** covariance matrix is used. (The **Prior Probabilities** setting is left at its default setting of **All groups equal**.) Finally, we request **Summary table** and **Leave-one-out classification** to evaluate the classification performance of our discriminant rule (see later).

Eigenvalues

Function	Eigenvalue	% of Variance	Cumulative %	Canonical Correlation
1	.930[a]	100.0	100.0	.694

a. First 1 canonical discriminant functions were used in the analysis.

Wilks' Lambda

Test of Function(s)	Wilks' Lambda	Chi-square	df	Sig.
1	.518	18.083	5	.003

Classification Function Coefficients

	place where skulls were found	
	Sikkem or Tibet	Lhasa
Greatest length of skull	1.468	1.558
Greatest horizontal breadth of skull	2.361	2.205
Height of skull	2.752	2.747
Upper face length	.775	.952
Face breadth between outermost points of cheek bones	.195	.372
(Constant)	-514.956	-545.419

Fisher's linear discriminant functions

Canonical Discriminant Function Coefficients

	Function 1
Greatest length of skull	.048
Greatest horizontal breadth of skull	-.083
Height of skull	-.003
Upper face length	.095
Face breadth between outermost points of cheek bones	.095
(Constant)	-16.222

Unstandardized coefficients

Display 12.6   Discriminant analysis output for skulls data.

The resulting discriminant analysis output is shown in Display 12.6. In the first part of this display, the eigenvalue (here 0.93) represents the ratio of the between-group sums of squares to the within-group sum of squares of the discriminant scores. It is this criterion that is maximized in discriminant function analysis (see Box 12.2). The *canonical correlation* is simply the Pearson correlation between the discriminant function scores and group membership coded as 0 and 1. For the skull data, the canonical correlation value is 0.694 so that $0.694 \times 0.694 \times 100 = 48.16\%$ of the

Functions at Group Centroids

place where skulls were found	Function 1
Sikkem or Tibet	-.877
Lhasa	.994

Unstandardized canonical discriminant functions evaluated at group means

Standardized Canonical Discriminant Function Coefficients

	Function 1
Greatest length of skull	.367
Greatest horizontal breadth of skull	-.578
Height of skull	-.017
Upper face length	.405
Face breadth between outermost points of cheek bones	.627

Display 12.6 (continued)

variance in the discriminant function scores can be explained by group differences.

The "Wilk's Lambda" part of Display 12.6 provides a test for assessing the null hypothesis that in the population the vectors of means of the five measurements are the same in the two groups (cf MANOVA in Chapter 5). The lambda coefficient is defined as the proportion of the total variance in the discriminant scores *not* explained by differences among the groups, here 51.84%. The formal test confirms that the sets of five mean skull measurements differ significantly between the two sites ($X^2(5) = 18.1$, $p = 0.003$). If the equality of mean vectors hypothesis had been accepted, there would be little point in carrying out a linear discriminant function analysis.

Next we come to the "Classification Function Coefficients". This table is displayed as a result of checking Fisher's in the Statistics sub-dialogue box (see Display 12.3). It can be used to find Fisher's linear discrimimant function as defined in Box 12.2 by simply subtracting the coefficients given for each variable in each group giving the following result:

$$z = -0.09 \times \text{meas1} + 0.156 \times \text{meas2} + 0.005 \times \text{meas3} - $$
$$0.177 \times \text{meas4} - 0.177 \times \text{meas5}$$

(12.1)

The difference between the constant coefficients provides the sample mean of the discriminant function scores

$$\bar{z} = 30.463 \qquad (12.2)$$

The coefficients defining Fisher's linear discriminant function in equation (12.1) are proportional to the unstandardized coefficients given in the "Canonical Discriminant Function Coefficients" table which is produced when **Unstandardized** is checked in the **Statistics** sub-dialogue box (see Display 12.3). The latter (scaled) discriminant function is used by SPSS to calculate *discriminant scores* for each object (skulls). The discriminant scores are centered so that they have a sample mean zero. These scores can be compared with the average of their group means (shown in the "Functions at Group Centroids" table) to allocate skulls into groups. Here the threshold against which a skull's discriminant score is evaluated is

$$0.0585 = 1/2 \times \left(-0.877 + 0.994\right) \qquad (12.3)$$

Thus new skulls with discriminant scores above 0.0585 would be assigned to the Lhasa site (type B); otherwise, they would be classified as type A.

When variables are measured on different scales, the magnitude of an unstandardized coefficient provides little indication of the relative contribution of the variable to the overall discrimination. The "Standardized Canonical Discriminant Function Coefficients" listed attempt to overcome this problem by rescaling of the variables to unit standard deviation. For our data, such standardization is not necessary since all skull measurements were in millimeters. Standardization should, however, not matter much since the within-group standard deviations were similar across different skull measures (see Display 12.4). According to the standardized coefficients, skull height (**meas3**) seems to contribute little to discriminating between the two types of skulls.

A question of some importance about a discriminant function is: how well does it perform? One possible method of evaluating performance is to apply the derived classification rule to the data set and calculate the misclassification rate. This is known as the *resubstitution estimate* and the corresponding results are shown in the "Original" part of the "Classification Results" table in Display 12.7. According to this estimate, 81.3% of skulls can be correctly classified as type A or type B on the basis of the discriminant rule. However, estimating misclassification rates in this way is known to be overly optimistic and several alternatives for estimating misclassification rates in discriminant analysis have been suggested (Hand, 1997). One of the most commonly used of these alternatives is the so-called *leaving one out method*, in which the discriminant function is first derived from only $n - 1$ sample members, and then used to classify the observation left out. The procedure is repeated n times, each time omitting

Prior Probabilities for Groups

place where skulls were found	Prior	Cases Used in Analysis	
		Unweighted	Weighted
Sikkem or Tibet	.500	17	17.000
Lhasa	.500	15	15.000
Total	1.000	32	32.000

Classification Results[b,c]

			Predicted Group Membership		
		place where skulls were found	Sikkem or Tibet	Lhasa	Total
Original	Count	Sikkem or Tibet	14	3	17
		Lhasa	3	12	15
	%	Sikkem or Tibet	82.4	17.6	100.0
		Lhasa	20.0	80.0	100.0
Cross-validated[a]	Count	Sikkem or Tibet	12	5	17
		Lhasa	6	9	15
	%	Sikkem or Tibet	70.6	29.4	100.0
		Lhasa	40.0	60.0	100.0

[a.] Cross validation is done only for those cases in the analysis. In cross validation, each case is classified by the functions derived from all cases other than that case.

[b.] 81.3% of original grouped cases correctly classified.

[c.] 65.6% of cross-validated grouped cases correctly classified.

Display 12.7   Classification output for skulls data.

a different observation. The "Cross-validated" part of the "Classification Results" table shows the results from applying this procedure (Display 12.7). The correct classification rate now drops to 65.6%, a considerably lower success rate than suggested by the simple resubstitution rule.

12.3.2 Tibetan Skulls: Uncovering Groups

We now turn to applying cluster analysis to the skull data. Here the prior classification of the skulls will be ignored and the data simply "explored" to see if there is any evidence of interesting "natural" groupings of the skulls and if there is, whether these groups correspond in anyway with Morant's classification.

Here we will use two hierarchical agglomerative clustering procedures, complete and average linkage clustering and then *k*-means clustering (see Box 12.1).

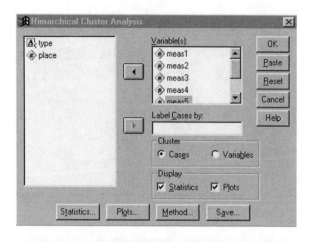

Display 12.8   Declaring a hierarchical cluster analysis.

A hierarchical cluster analysis can be requested in SPSS using the commands

Analyze – Classify – Hierarchical Cluster...

This opens the Hierarchical Cluster Analysis dialogue box where the variables to be used in the cluster analysis are specified under the Variable(s) list (Display 12.8). Our interest is in clustering cases (skulls) rather than variables and so we keep Cases checked under the Cluster part of the box. By checking the Statistics.... button and the Plots... button in this dialogue box, further boxes appear (Displays 12.9 and 12.10) that allow particular tables and plots to be chosen.

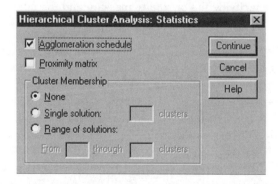

Display 12.9   Specifying table output of a hierarchical cluster analysis.

Display 12.10   Requesting a dendrogram.

Display 12.11   Specifying the proximity measure and the linking (cluster) method.

The **Method...** button opens the **Method** sub-dialogue box and determines the cluster method to be employed (Display 12.11). Hierarchical clustering involves, essentially, two main steps: First, the data is converted into a proximity matrix the elements of which are either distances or similarities between each pair of skulls. These are then used to combine skulls according to the process outlined in Box 12.2.

SPSS offers a variety of proximity measures under the **Measure** part of the **Method** sub-dialogue box. The measures are listed according to the type of data for which they can be used (for more details on proximity measures, see Everitt et al., 2001). All our five variables are measured on an interval scale making Euclidean distance an obvious choice (Display 12.11).

Agglomeration Schedule

Stage	Cluster Combined Cluster 1	Cluster Combined Cluster 2	Coefficients	Stage Cluster First Appears Cluster 1	Stage Cluster First Appears Cluster 2	Next Stage
1	8	13	3.041	0	0	4
2	15	17	5.385	0	0	14
3	9	23	5.701	0	0	11
4	8	19	5.979	1	0	8
5	24	28	6.819	0	0	17
6	21	22	6.910	0	0	21
7	16	29	7.211	0	0	15
8	7	8	8.703	0	4	13
9	2	3	8.874	0	0	14
10	27	30	9.247	0	0	23
11	5	9	9.579	0	3	13
12	18	32	9.874	0	0	18
13	5	7	10.700	11	8	24
14	2	15	11.522	9	2	28
15	6	16	12.104	0	7	22
16	14	25	12.339	0	0	21
17	24	31	13.528	5	0	23
18	11	18	13.537	0	12	22
19	1	20	13.802	0	0	26
20	4	10	14.062	0	0	28
21	14	21	15.588	16	6	25
22	6	11	16.302	15	18	24
23	24	27	18.554	17	10	27
24	5	6	18.828	13	22	29
25	12	14	20.700	0	21	30
26	1	26	24.597	19	0	27
27	1	24	25.269	26	23	30
28	2	4	25.880	14	20	29
29	2	5	26.930	28	24	31
30	1	12	36.342	27	25	31
31	1	2	48.816	30	29	0

Display 12.12   Complete linkage output for skulls data.

(Since all five measurements are in millimeters and each has similar standard deviations — see Display 12.4 — we will not standardize them before calculating distances.)

The hierarchical (linkage) method can be chosen from the Cluster Method drop-down list. SPSS provides all the standard linkage methods, including complete and average linkage. To start with, we request complete linkage clustering by selecting Furthest neighbor (furthest neighbor is a synonym for complete linkage).

The complete linkage clustering output is shown in Display 12.12. The columns under the "Cluster Combined" part of the "Agglomeration Schedule" table shows which skulls or clusters are combined at each stage of the cluster procedure. First, skull 8 is joined with skull 13 since the

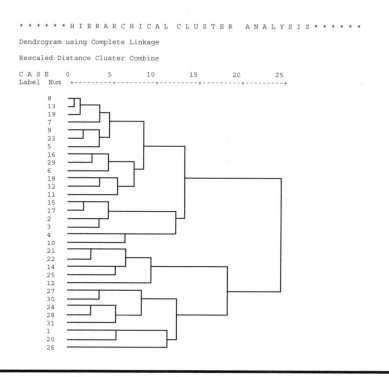

Display 12.12 (continued)

Euclidean distance between these two skulls is smaller than the distance between any other pair of skulls. The distance is shown in the column labeled "Coefficients". Second, skull 15 is joined with skull 17 and so on. SPSS uses the number of the first skull in a cluster to label the cluster. For example, in the fourth step, skull 19 is joined with the cluster consisting of skull 8 and skull 13 (labeled "cluster 8" at this stage). When clusters are joined, the "Coefficient" value depends on the linkage method used. Here with complete linkage, the distance between "cluster 8" and skull 19 is 5.98 millimeters since this is the largest distance between skull 19 and any of the members of "cluster 8" (distance to skull 13).

The columns under the heading "Stage Cluster First Appears" show the stage at which a cluster or skull being joined first occurred in its current form. For example, "cluster 8" joined at stage 4 was constructed at stage 1. Finally, the "Next Stage" column shows when a cluster constructed at the current stage will be involved in another joining. For example, "cluster 8" as constructed in stage 4 (skulls 8, 13, and 19) will not be used until stage 8 where it is linked with skull 7.

It is often easier to follow how groups and individual skulls join together in this process in a *dendrogram*, which is simply a tree-like

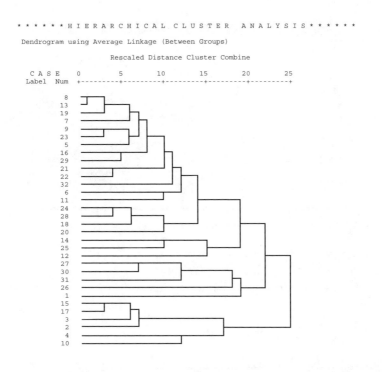

Display 12.13 Dendrogram from average linkage clustering of skulls data.

diagram that displays the series of fusions as the clustering proceeds for individual sample members to a single group. The dendrogram in this case is shown in the output in Display 12.12. (However, note that SPSS rescales the joining distances to the range 0 to 25.)

The dendrogram may, on occasions, also be useful in deciding the number of clusters in a data set with a sudden increase in the size of the difference in adjacent steps taken as an informal indication of the appropriate number of clusters to consider. For the dendrogram in Display 12.12, a fairly large jump occurs between stages 29 and 30 (indicating a three-group solution) and an even bigger one between this penultimate and the ultimate fusion of groups (a two-group solution).

To now apply average linkage clustering based on Euclidean distances between pairs of skulls, we need to select **Between-groups linkage** as the **Cluster Method** in the **Method** sub-dialogue box (see Display 12.11). The resulting dendrogram is shown in Display 12.13. The initial steps agree with the complete linkage solution, but eventually the trees diverge with the average linkage dendrogram successively adding small clusters to one increasingly large cluster. (Note that the joining distances are not comparable across different dendograms due to differential rescaling.) For the

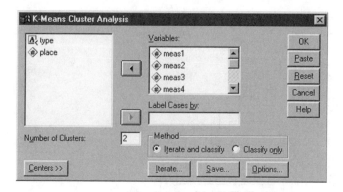

Display 12.14   Declaring a k-means cluster analysis.

average linkage dendrogram in Display 12.13 it is not clear where to cut the dendrogram to give a specific number of groups.

Finally, we can apply the *k*-means clustering procedure by using the commands

Analyze – Classify – K-Means Cluster...

This opens the **K-Means Cluster Analysis** dialogue box where the variables used in the clustering need to be specified (Display 12.14). The number of clusters needs to be set by the investigator. Here we request a two-group cluster solution since the complete linkage dendrogram has suggested the possibility of a "natural" partition into two sets of skulls.

The resulting cluster output is shown in Display 12.15. The "Initial Cluster Centers" table shows the starting values used by the algorithm (see Box 12.1). The "Iteration History" table indicates that the algorithm has converged and the "Final Cluster Centers" and "Number of Cases in each Cluster" tables describe the final cluster solution. In *k*-means clustering, the cluster centers are determined by the means of the variables across skulls belonging to the cluster (also known as the *centroids*). As is typical for a two-group partition based on positively correlated variables, the centroids indicate a cluster with higher scores on all of the five measurement variables (cluster 1, containing 13 skulls) and one with lower scores (cluster 2, containing 19 skulls). We also see that the clusters differ most on **meas1** (greatest length of skulls), **meas5** (face breadth between outermost points of cheek bones), and **meas4** (upper face length).

How does the *k*-means two-group solution compare with Morant's original classification of the skulls into types A and B? We can investigate this by first using the **Save...** button on the **K-Means Cluster Analysis** dialogue box to save cluster membership for each skull in the **Data View** spreadsheet. The new categorical variable now available (labeled qcl_1) can be cross-tabulated

Initial Cluster Centers

	Cluster	
	1	2
Greatest length of skull	200.0	167.0
Greatest horizontal breadth of skull	139.5	130.0
Height of skull	143.5	125.5
Upper face length	82.5	69.5
Face breadth between outermost points of cheek bones	146.0	119.5

Iteration History[a]

	Change in Cluster Centers	
Iteration	1	2
1	16.626	16.262
2	.000	.000

a. Convergence achieved due to no or small distance change. The maximum distance by which any center has changed is .000. The current iteration is 2. The minimum distance between initial centers is 48.729.

Final Cluster Centers

	Cluster	
	1	2
Greatest length of skull	188.4	174.1
Greatest horizontal breadth of skull	141.3	137.6
Height of skull	135.8	131.6
Upper face length	77.6	69.7
Face breadth between outermost points of cheek bones	138.5	130.4

Number of Cases in each Cluster

Cluster	1	13.000
	2	19.000
Valid		32.000
Missing		.000

Display 12.15   k-means cluster output for skulls data.

with assumed skull type (variable **place**). Display 12.16 shows the resulting table; the *k*-means clusters largely agree with the skull types as originally suggested by Morant, with cluster 1 consisting primarily of Type B skulls (those from Lhasa) and cluster 2 containing mostly skulls of Type A (from Sikkim and the neighboring area of Tibet). Only six skulls are wrongly placed.

Cluster Number of Case * place where skulls were found Crosstabulation

				place where skulls were found		Total
				Sikkem or Tibet	Lhasa	
Cluster Number of Case	1	Count		2	11	13
		% within place where skulls were found		11.8%	73.3%	40.6%
	2	Count		15	4	19
		% within place where skulls were found		88.2%	26.7%	59.4%
Total		Count		17	15	32
		% within place where skulls were found		100.0%	100.0%	100.0%

Display 12.16   Cross-tabulation of cluster-membership by site.

12.4 Exercises

12.4.1 Sudden Infant Death Syndrome (SIDS)

The data below were collected by Spicer et al. (1987) in an investigation of sudden infant death syndrome (SIDS). Data are given for 16 SIDS victims and 49 controls. All children had a gestational age of 37 or more weeks and were regarded as full term. Four variables were recorded for each infant as follows:

- Heart rate in beats per minute (HR)
- Birthweight in grams (BW)
- A measurement of heart and lung function (Factor68)
- Gestational age in weeks (Gesage)

Also given is the group variable, 1 = control, 2 = SIDS.

1. Using only the birthweight and heart and lung function measurement, construct a scatterplot of the data that shows the location of Fisher's linear discriminant function and labels the infants by group. (Hint: save the group membership predicted for each child according to the discriminant rule.)
2. Find Fisher's linear discriminant function based on all four variables and determine into which group this would assign an infant with the following four measurements:

   HR = 120.0, BW = 2900, Factor68 = 0.500, Gesage = 39

Table 12.2   SIDS Data

Group	HR	BW	Factor68	Gesage
1	115.6	3060	0.291	39
1	108.2	3570	0.277	40
1	114.2	3950	0.39	41
1	118.8	3480	0.339	40
1	76.9	3370	0.248	39
1	132.6	3260	0.342	40
1	107.7	4420	0.31	42
1	118.2	3560	0.22	40
1	126.6	3290	0.233	38
1	138	3010	0.309	40
1	127	3180	0.355	40
1	127.7	3950	0.309	40
1	106.8	3400	0.25	40
1	142.1	2410	0.368	38
1	91.5	2890	0.223	42
1	151.1	4030	0.364	40
1	127.1	3770	0.335	42
1	134.3	2680	0.356	40
1	114.9	3370	0.374	41
1	118.1	3370	0.152	40
1	122	3270	0.356	40
1	167	3520	0.394	41
1	107.9	3340	0.25	41
1	134.6	3940	0.422	41
1	137.7	3350	0.409	40
1	112.8	3350	0.241	39
1	131.3	3000	0.312	40
1	132.7	3960	0.196	40
1	148.1	3490	0.266	40
1	118.9	2640	0.31	39
1	133.7	3630	0.351	40
1	141	2680	0.42	38
1	134.1	3580	0.366	40
1	135.5	3800	0.503	39
1	148.6	3350	0.272	40
1	147.9	3030	0.291	40
1	162	3940	0.308	42
1	146.8	4080	0.235	40
1	131.7	3520	0.287	40
1	149	3630	0.456	40

Table 12.2 (continued)   SIDS Data

Group	HR	BW	Factor68	Gesage
1	114.1	3290	0.284	40
1	129.2	3180	0.239	40
1	144.2	3580	0.191	40
1	148.1	3060	0.334	40
1	108.2	3000	0.321	37
1	131.1	4310	0.45	40
1	129.7	3975	0.244	40
1	142	3000	0.173	40
1	145.5	3940	0.304	41
2	139.7	3740	0.409	40
2	121.3	3005	0.626	38
2	131.4	4790	0.383	40
2	152.8	1890	0.432	38
2	125.6	2920	0.347	40
2	139.5	2810	0.493	39
2	117.2	3490	0.521	38
2	131.5	3030	0.343	37
2	137.3	2000	0.359	41
2	140.9	3770	0.349	40
2	139.5	2350	0.279	40
2	128.4	2780	0.409	39
2	154.2	2980	0.388	40
2	140.7	2120	0.372	38
2	105.5	2700	0.314	39
2	121.7	3060	0.405	41

12.4.2 Nutrients in Food Data

The data given below show the nutrients in different foodstuffs. (The quantity involved is always three ounces.) Use a variety of clustering techniques to discover if there is any evidence of different groups of foodstuffs and whether or not the different techniques you use produce the same (or very similar) solutions.

12.4.3 More on Tibetan Skulls

1. The results given in the chapter show that the Type A and Type B skulls can be discriminated using the five measurements available, but do all five variables contribute to this discrimination?

Table 12.3   Nutrients in Foodstuffs

		Food Energy (Calories)	Protein (Grams)	Fat (Grams)	Calcium (Milligrams)	Iron (Milligrams)
BB	Beef, braised	340	20	28	9	2.6
HR	Hamburger	245	21	17	9	2.7
BR	Beef, roast	420	15	39	7	2.0
BS	Beef, steak	375	19	32	9	2.5
BC	Beef, canned	180	22	10	17	3.7
CB	Chicken, broiled	115	20	3	8	1.4
CC	Chicken, canned	170	25	7	12	1.5
BH	Beef heart	160	26	5	14	5.9
LL	Lamb leg, roast	265	20	20	9	2.6
LS	Lamb shoulder, roast	300	18	25	9	2.3
HS	Smoked ham	340	20	28	9	2.5
PR	Pork, roast	340	19	29	9	2.5
PS	Pork, simmered	355	19	30	9	2.4
BT	Beef tongue	205	18	14	7	2.5
VC	Veal cutlet	185	23	9	9	2.7
FB	Bluefish, baked	135	22	4	25	0.6
AR	Clams, raw	70	11	1	82	6.0
AC	Clams, canned	45	7	1	74	5.4
TC	Crabmeat, canned	90	14	2	38	0.8
HF	Haddock, fried	135	16	5	15	0.5
MB	Mackerel, broiled	200	19	13	5	1.0
MC	Mackerel, canned	155	16	9	157	1.8
PF	Perch, fried	195	16	11	14	1.3
SC	Salmon, canned	120	17	5	159	0.7
DC	Sardines, canned	180	22	9	367	2.5
UC	Tuna, canned	170	25	7	7	1.2
RC	Shrimp, canned	110	23	1	98	2.6

Investigate using the stepwise variable selection procedures available in SPSS.2.

2. Save the discriminant function scores for each skull and then construct a histogram of the scores. Comment on the shape of the histogram.

3. Compare the two-group solution from complete and average linkage clustering to both the two-group solution from *k*-means clustering and the Type A/Type B classification. Comment on the results. Display each solution and the original classification of the skulls in the space of the first two principal components of the data.

References

Agresti, A. (1996) *Introduction to Categorical Data Analysis*. New York: Wiley.

Aitkin, M. (1978) The analysis of unbalanced cross-classifications (with discussion). *Journal of the Royal Statistical Society, A*, 41, 195-223.

Altman, D. G. (1991) *Practical Statistics for Medical Research*. London: Chapman and Hall.

Altman, D. G. (1998) Categorizing continuous variables. In *Encyclopedia of Biostatistics Volume 1* (P. Armitage and T. Colton, Eds.). Chichester: Wiley.

Beck, A. T., Steer, A., and Brown G. K. (1996) *Beck Depression Inventory Manual* (2nd ed). San Antonio: The Psychological Corporation.

Belsley, Kuh, and Welsh (1980). *Regression Diagnostics: Identifying Influential Data and Sources of Collinearity*. New York: Wiley.

Berger, R.L., Boos, D.D., and Guess, F.M. (1988) Tests and confidence sets for comparing two mean residual life functions. *Biometrics*, 44, 103–115.

Box, G. E. P. (1954) Some theorems on quadratic forms applied in the study of analysis of variance problems. II. Effects of inequality of variance and of correlations between errors in the two-way classification. *Annals of Mathematical Statistics*, 25, 484–498.

Breslow, N. E. (1970) A generalized Kruskal-Wallace test for comparing K samples subject to unequal patterns of censorship. *Biometrika*, 57, 579–594.

Cleveland, W. S. (1979) Robust locally weighted regression and smoothing scatterplots. *Journal of the American Statistical Association*, 74, 829–836.

Cleveland, W. S. (1993) *Visualizing Data*. Summit, NJ: Hobart.

Collett, D. (2003) *Modelling Survival Data in Medical Research* (2nd ed). Boca Raton, FL: Chapman and Hall/CRC.

Collett, D. (2003) *Modeling Binary Data* (2nd ed). Boca Raton, FL: Chapman and Hall/CRC.

Conover, W. J. (1998) *Practical Nonparametric Statistics*. New York: John Wiley & Sons.

Cook, R. D. (1977) Detection of influential observations in linear regression. *Technometrics*, 19, 15–18.

Cook, R. D. and Weisberg, S. (1982) *Residuals and Influence in Regression*. London: Chapman and Hall.

Cox, D. R. (1972) Regression models and life tables. *Journal of the Royal Statistical Society, B.*, 34, 187–220.

Cox, D. R. and Snell, E. J. (1968) A general definition of residuals. *Journal of the Royal Statistical Society, B*, 30, 248-275.

Davidson, M. L. (1972) Univariate versus multivariate tests in repeated measures experiments. *Psychological Bulletin*, 77, 446–452.

Der, G. and Everitt, B. S. (2001) *A Handbook of Statistical Analysis Using SAS* (2nd ed). Boca Raton, FL: Chapman and Hall/CRC.

Diggle, P. J. (1988) An approach to the analysis of repeated measures. *Biometrics*, 44, 959–971.

Dizney, H. and Gromen, L. (1967) Predictive validity and differential achievement on three MLA comparative foreign language tests. *Educational and Psychological Measurement*, 27, 1959-1980.

Draper, N. R. and Smith, H. (1998) *Applied Regression Analysis* (3rd ed). New York: Wiley.

Dunn, G. and Everitt, B. S. (1995) *Clinical Biostatistics: An Introduction to Evidence-Based Medicine*. London: Arnold.

Everitt, B. S. (1992) *The Analysis of Contingency Tables* (2nd ed). Boca Raton, FL: Chapman and Hall/CRC.

Everitt, B. S. (1999) *Chance Rules*. New York: Springer Verlag.

Everitt, B. S. (2001a) *A Handbook of Statistical Analysis Using S-PLUS* (2nd ed). Boca Raton, FL: Chapman and Hall/CRC.

Everitt, B. S. (2001b) *Statistics for Psychologists*. Mahwah, NJ: Lawrence Erlbaum.

Everitt, B. S. (2002a) *The Cambridge Dictionary of Statistics* (2nd ed). Cambridge: Cambridge University Press.

Everitt, B. S. (2002b) *Modern Medical Statistics: A Practical Guide*. London: Arnold.

Everitt, B. S. and Dunn, G. (2001) *Applied Multivariate Data Analysis* (2nd ed). London: Arnold.

Everitt, B. S. and Hay, D. (1992) *Talking About Statistics*. London: Arnold.

Everitt, B. S., Landau, S., and Leese, M. (2001) *Cluster Analysis* (4th ed). London: Arnold.

Everitt, B. S. and Pickles, A. (2000) *Statistical Aspects of the Design and Analysis of Clinical Trials*. London: ICP.

Everitt, B. S. and Rabe-Hesketh, S. (2001) *Analysing Medical Data Using S-PLUS*. New York: Springer.

Everitt, B. S. and Wykes, T. (1999) *A Dictionary of Statistics for Psychologists*. London: Arnold.

Fisher, R. A. (1936) The use of multiple measurements on taxonomic problems. *Annals of Eugenics*, 7, 179–188.

Gamerman, D. (1991) Dynamic Bayesian models for survival data. *Applied Statistics*, 40, 63-79.

Goldberg, D. (1972) *The Detection of Psychiatric Illness by Questionnaire*. Oxford: Oxford University Press.

Greenhouse, S. W. and Geisser, S. (1959) On the methods in the analysis of profile data. *Psychometrika*, 24, 95-112.

Hand, D. J. (1997) *Construction and Assessment of Classification Rules*. Chichester: Wiley.

Hand, D. J. (1998) Discriminant Analysis, Linear. In *Encyclopedia of Biostatistics Volume 2* (P. Armitage and T. Colton, Eds.). Chichester: Wiley.

Hand, D. J., Daly, F., Lunn, A. D., et al. (1994) *Small Data Sets*. Boca Raton, FL: Chapman and Hall/CRC.

Harrell, E. E. (2000) *Regression Modelling Strategies with Applications to Linear Models, Logistic Regression and Survival Analysis*. New York: Springer.

Hollander, M. and Wolfe, D. A. (1999) *Nonparametric Statistical Methods*. New York: Wiley.

Hosmer, D. W. and Lemeshow, S. (1989) *Applied Logistic Regression* (2nd ed). New York: Wiley.

Hosmer, D. W. and Lemeshow, S. (1999) *Applied Survival Analysis*. New York: Wiley.

Howell, D. C. (2002) *Statistical Methods for Psychology* (5th ed). Pacific Lodge, CA: Duxbury.

Huberty, C. J. (1994) *Applied Discriminant Analysis*. New York: Wiley.

Huynh, H. and Feldt, L. S. (1976) Estimates of the correction for degrees of freedom for sample data in randomised block and split-plot designs. *Journal of Educational Statistics*, 1, 69–82.

Jolliffe, I. T. (2002) *Principal Components Analysis* (2nd ed). New York: Springer.

Kapor, M. (1981) *Efficiency on Erogocycle in Relation to Knee-Joint Angle and Drag*. Delhi: University of Delhi.

Kleinbaum, D. G. and Klein, M. (2002) *Logistic Regression — A Self Learning Text*. New York: Springer.

Krzanowski, W. J. and Marriott, F. H. C. (1995) *Multivariate Analysis Part 2*. London: Arnold.

Levene, H. (1960a) Robust tests for the equality of variance. In *Contributions to Probability and Statistics* (O. Aikin, Ed.). Stanford, CA: Stanford University Press.

Manly (1999). *Randomization, Bootstrap, and Monte Carlo Methods in Biology*. Boca Raton, FL: Chapman and Hall/CRC.

Mardia, K. V., Kent, J. T., and Bibby, J. M. (1979) *Multivariate Analysis*. London: Academic Press.

Marriott, F. H. C. (1974) *The Interpretation of Multiple Observations*. London: Academic Press.

Maxwell, S. E. and Delaney, H. D. (1990) *Designing Experiments and Analyzing Data*. Stamford, CT: Wadsworth.

McCullagh, P. and Nelder, J. A. (1989) *Generalized Linear Models* (2nd ed). Boca Raton, FL: Chapman and Hall/CRC.

McKay, R. J. and Campbell, N. A. (1982(a)) Variable selection techniques in discriminant analysis. I. Description. *British Journal of Mathematical and Statistical Psychology*, 35, 1–29.

McKay, R. J. and Campbell, N. A. (1982(b)) Variable selection techniques in discriminant analysis. II. Allocation. *British Journal of Mathematical and Statistical Psychology*, 35, 30–41.

Miles, J. and Shevlin, M. (2001) *Applying Regression and Correlation*. London: Sage Publications.

Morant, G. M. (1923) A first study of the Tibetan skull. *Biometrika*, 14, 193–260.

Nelder, J. A. (1977) A reformulation of linear models. *Journal of the Royal Statistical Society, A*, 140, 48–63.

Novince, L. (1977) The contribution of cognitive restructuring to the effectiveness of behavior rehearsal in modifying social inhibition in females. Cincinnati, OH: University of Cininnati.

Pagano, R. R. (1990) *Understanding Statistics in the Behavioral Sciences* (3rd ed). St Paul, MN: West Publishing Co.

Pagano, R. R. (1998) *Understanding Statistics in the Behavioral Sciences* (5th ed). Stamford, CT: Wadsworth.

Piexoto, J. L. (1990) A property of well-formulated polynomial regression models. *American Statistician*, 44, 26-30.

Preacher, K. J. and MacCallum, R. C. (2003) Repairing Tom Swift's Electric Factor Analysis. *Understanding Statistics*, 2, 13–44.

Proudfoot, J., Goldberg, D., Mann, A., Everitt, B. S., Marks, I. M., and Gray, J. A. (2003) Computerised, interactive, multimedia cognitive behavioural therapy for anxiety and depression in general practice. *Psychological Medicine,* 33, 217–228.

Rabe-Hesketh, S. and Skrondal, A. (2003) *Generalized Latent Variable Modeling: Multilevel, Longitudinal, and Structural Equation Models*. Boca Raton, FL: Chapman and Hall/CRC.

Rawlings, J. O., Pantula, S. G., and Dickey, A. D. (1998) *Applied Regression Analysis*. New York: Springer.

Rossman, A. (1996) *Workshop Statistics: Discovery with Data*. New York: Springer Verlag.

Rothschild, A. J., Schatzberg, A. F., Rosenbaum, A. H., et al. (1982) The dexamethasone suppression test as a discriminator among subtypes of psychotic patients. *British Journal of Psychiatry*, 141, 471–474.

Sartwell, P. E., Mazi, A. T., Aertles, F. G., et al. (1969) Thromboembolism and oral contraceptives: an epidemiological case-control-study. *American Journal of Epidemiology*, 90, 365-375.

Schmidt, U., Evans, K., Tiller, J., and Treasure, J. (1995) Puberty, sexual milestones and abuse: How are they related in eating disorder patients? *Psychological Medicine,* 25, 413–417.

Schoenfeld, D. A. (1982) Partial residuals for the proportional hazards regression model. *Biometrika*, 39, 499–503.

Spicer, C. C., Laurence, G. J., and Southall, D. P. (1987) Statistical analysis of heart rates and subsequent victims of sudden infant death syndrome. *Statistics in Medicine*, 6, 159–166.

SPSS Inc. (2001a) *SPSS 11.0 Advanced Models*: Englewood Cliffs, NJ: Prentice Hall.

SPSS Inc. (2001b) *SPSS 11.0 Regression Models*: Englewood Cliffs, NJ: Prentice Hall.

SPSS Inc. (2001c) *SPSS 11.0 Syntax Reference Guide*: Englewood Cliffs, NJ:Prentice Hall.

SPSS Inc. (2001d) *SPSS Base 11.0 for Windows User's Guide*: Englewood Cliffs, NJ:Prentice Hall.

Stevens, J. (1992) *Applied Multivariate Statistics for the Social Sciences,* Hillsdale, NJ: Erlbaum.

Tarone, R. E. and Ware, J. (1977) On distribution free tests for equality of survival distributions. *Biometrika,* 64, 156–160.

Therneau, T. M., Grambsch, P. M., and Fleming, T. R. (1990) Martingale-based residuals for survival models. *Biometrika,* 77, 147–160.

Therneau, T. M. and Grambsch, P. M. (2000) *Modeling Survival Data.* New York: Springer.

Wechsler, D. (1974) *Wechsler Intelligence Scale for Children — Revised.* New York: Psychological Corp.

Witkin, H. A., Oftman, P. K., Raskin, E., and Karp, S. A. (1971) *Group Embedded Figures Test Manual,* Palo Alto, CA: Consulting Psychologist Press.

Index

A

Add Cases dialogue box, 14
Add option, 178
Add Variables dialogue box, 14
Advanced Models add-on module, 1, 2
Afterlife belief, 61, 62, 74–75
Age by gender interaction, 241
Age_cat variable, 227
Ages at marriage (husbands and wives)
    categorical data, 68–71
    continuous data, 27, 29, 43–56, 60
Agesex variable, 253, 254, 256
Age variable
    basics, 4
    examples, 15, 16
    logistic regression, 224, 227
    multiple linear regression, 94, 96
    survival analysis, 253, 261, 265, 267
Agglomeration Schedule table, 321
Agglomerative hierarchical techniques, 308
Aggregate menu selection, 14
Agresti studies, 61, 62, 67, 84, 87, 101
AIDS patients' physician evaluations, 277,
        279, 280–281, 293–302, 304
Air pollution (United States)
    multiple linear regression, 123–125
    principal components and factor
        analyses, 302, 303
Aitkin studies, 165
AIX platform, 1
Alcohol

categorical data, 62–63, 81–82
    linear mixed effects models, 216
Align variable, 8
All command, 26
All Groups Equal option, 314
Altman studies, 28, 68
Ampersand, 17
Analysis, see also Factorial designs; One-
        way designs
    categorical data, 63–86
    continuous data, 33–56
    multiple linear regression, 94–123
Analysis of covariance (ANCOVA)
    factorial designs, 168–169
    repeated measures, 207
Analysis of variance (ANOVA), see also
        Multivariate analysis of variance
        (MANOVA)
    one-way designs, 129–133, 143, 144
    two-way designs, 152–155, 162
Analysis of variance (ANOVA), repeated
        measures
    ANOVA approach, 173–175
    blood glucose levels, 186, 187
    correction factors, 174–175
    data description, 171
    exercises, 185–187
    field dependence, 171, 172
    lens strength, 185, 186
    multivariate analysis of variance, 175,
        182

DATE DUE

GAYLORD